Science, Technology and Innovation Studies

Series Editors

Leonid Gokhberg
Moscow, Russia

Dirk Meissner
Moscow, Russia

Science, technology and innovation (STI) studies are interrelated, as are STI policies and policy studies. This series of books aims to contribute to improved understanding of these interrelations. Their importance has become more widely recognized, as the role of innovation in driving economic development and fostering societal welfare has become almost conventional wisdom. Interdisciplinary in coverage, the series focuses on the links between STI, business, and the broader economy and society. The series includes conceptual and empirical contributions, which aim to extend our theoretical grasp while offering practical relevance. Relevant topics include the economic and social impacts of STI, STI policy design and implementation, technology and innovation management, entrepreneurship (and related policies), foresight studies, and analysis of emerging technologies. The series is addressed to professionals in research and teaching, consultancies and industry, government and international organizations.

More information about this series at http://www.springer.com/series/13398

Fred Phillips

What About the Future?

New Perspectives on Planning, Forecasting and Complexity

 Springer

Fred Phillips
Anderson School of Management
University of New Mexico
Albuquerque, NM, USA

ISSN 2570-1509   ISSN 2570-1517   (electronic)
Science, Technology and Innovation Studies
ISBN 978-3-030-26167-2   ISBN 978-3-030-26165-8   (eBook)
https://doi.org/10.1007/978-3-030-26165-8

This Springer imprint is published by the registered company Springer Nature Switzerland AG.
The registered company address is: Gewerbestrasse 11, 6330 Cham, Switzerland

I dedicate this book to the memory of my mentor Hal Linstone (1924–2016). I refer readers to Hal's Remembrance, in Technological Forecasting and Social Change, *volume 111, page 1, 2016.*

Preface

As Editor-in-Chief of the journal *Technological Forecasting and Social Change*, I've accumulated what seems to be a lot of books about the future. I can't complain. As so little is known about the future, how much can be written about it? Surely my colleagues in the history department are burdened with many, many more books!

Anyway, books on the future seem to fall into two kinds. The first kind makes specific predictions of social and technological developments and tells companies what they should do about them. The second kind are textbooks and manuals on forecasting techniques. Their topic coverage, it seems to me, does not provide what modern readers need.

Readers need to know how to form a fundamental perspective about the future—before they start to try to predict things.

Who are these readers? Students and graduates, whether they find themselves in the workplace yet or not, need such a perspective. (Or more accurately, the skills to form their own perspective. These skills are rarely taught in schools or universities.) Students taking a first course in strategy or public planning need this guidance for professional as well as personal reasons. As futurist Peter Bishop remarked, "It is critical to empower young people by showing them how to anticipate a range of possibilities and to influence the course of events."

Researchers, corporate strategists, practicing futurists, and government and NGO officials engaging in foresight exercises should occasionally "go back to basics," to firm up the foundations of their work. The ideas herein should help them in that regard.

I have assumed that all readers of this book work or study in an organization of some kind, or are about to, and are concerned about organizational, political, and personal/family futures. This book's examples move from each of those domains to another, sometimes abruptly, though the jumps should not be hard to follow.

Yet the book's main focus is on the individual. Readers will work for, or start, many companies during their working life. The average number of years any firm stays in the Fortune 500 continues to drop. I won't try to buck that tide by giving advice to companies. People, not firms, need this book's ideas. If they can then help their companies, so much the better.

I've written in a casual but precise style that should make the book easily digestible yet practically useful for all. I hope, also, to convince readers that the often simple perspectives presented here have serious and far-reaching implications.

Students and researchers may appreciate the footnoted citations. All readers are free to ignore the footnotes and just enjoy the text. Helpful "takeaways" at the end of each chapter summarize the most useful bits.

Some passages are updates, condensations, or expansions of material published earlier in scattered form, in my 2001 book *Market-Oriented Technology Management: Innovating for Profit in Entrepreneurial Times,* in my blogs, or in articles I've written for *Technological Forecasting and Social Change,* herein brought together in a more useful format. Most of this book, however, is newly written. I hope it fills the gap between what's in currently available books (on the one hand) and what readers need (on the other).

The book reflects the "multiple perspectives" and systems-thinking viewpoints pioneered in *TF&SC* by Dr. Harold Linstone, founding editor of the journal. I resonated with these viewpoints before becoming involved with the journal, extended them together with Hal during the years we worked together on the journal, and continued to embrace them when Hal handed the reins of the journal to me.

Albuquerque, NM   Fred Phillips
June 2019

Acknowledgments

My grandfather, Leon Julius Brandt, was a printer by trade. He painstakingly set letter type and engraved plates, by hand. This plate of his eerily asks, "What about the future?" As I desktop-typeset this book on Microsoft Word for Mac, changing fonts at will and in seconds, I say to his ghost, "Well, Leon, this is it!"

My father, Herbert Phillips, an inventor who also took engineering into the public policy arena, inspired my own approach to career and to this book. I also wish to acknowledge the nearly 100 members of the *Technological Forecasting and Social Change* editorial board, from whom I've learned so much and whose ideas, directly or indirectly, are reflected in this book. In particular, Leonid Gokhberg and Dirk Meissner of the Higher School of Economics in Moscow, editors of the *Science, Technology and Innovation Studies* series at Springer, initially spurred the idea for this book.

Thanks as well to my research assistant, La Anh Alice Nguyen, to illustrator Lily Nguyen, and to the institutions that have supported my writing time: Yuan Ze and National Chengchi Universities in Taiwan; University of New Mexico, USA; and the Institute of Geographic Sciences and Natural Resources Research at the Chinese Academy of Sciences (CAS) in China. Peng Lu, Guo Baishu, and Jin Gui of CAS helped with final details.

Excerpts from F. Phillips, "Triple Helix and the Circle of Innovation." *Journal of Contemporary Eastern Asia* (JCEA). Vol. 13, No. 1, April/May 2014, 57-68, are presented here under the journal's Creative Commons license.

Excerpts from the following published articles appear here by kind permission of Elsevier:

- F. Phillips, "Change in socio-technical systems: Researching the Multis, the Biggers, and the More Connecteds." *Technological Forecasting & Social Change*, 75(5), June 2008, 721–734.
- F. Phillips and H. Linstone, "Key Ideas from a 25-Year Collaboration at TFSC." *Technological Forecasting & Social Change,* Vol. 105, pp.158–166, April, 2016.
- F. Phillips, "The Globalization Paradox." *Technological Forecasting & Social Change*, 2018. https://doi.org/10.1016/j.techfore.2018.06.016

Excerpts from F. Phillips, "Perspectives on Big Data." *Science and Public Policy,* 44(5) 2017, 730–737, appear here by permission of Oxford University Press.

Contents

About the Author

Fred Phillips is Editor-in-Chief of the international journal *Technological Forecasting and Social Change.* He is currently a Professor at the University of New Mexico and Visiting Scientist at the Chinese Academy of Sciences in Beijing. He is the 2017 winner of the Kondratieff Medal, awarded by the Russian Academy of Sciences, and 2019 recipient of the INEKA Medal. He is a Senior Fellow (and formerly Research Director) at the IC^2 Institute of the University of Texas at Austin and a PICMET Fellow. He is a partner in General Informatics, LLC, providing advisory services to corporations, governments, and international agencies on research policy and technology-based economic development.

What Do We Mean by "The Future"?

The future ain't what it used to be.
—Yogi Berra (It is popularly thought that Einstein and Berra
said these things. We're not sure. See
https://quoteinvestigator.com/2012/12/06/future-not-used/)

Does the future exist? We cannot see it. We might doubt the reality of other unseen entities—germs? gods? Higgs bosons?—so why not doubt the existence of the future? The answer is that unlike other unseen things, the future continuously makes parts of itself visible. Each tomorrow has reliably turned into today, every 24 h since you and I were born—and reportedly, long before that!

Albert Einstein said, "I never think about the future; it comes soon enough."[1] Gurus advise us to "live in the present." Following their advice, undeniably, calms our minds. Yet we cannot ignore the future. It will arrive. Some of it during our lifetimes, more of it in our children's lifetimes, and in their children's.

Yogi was right: The future is more complicated than it used to be. Our parents urged us to think about our futures: "Get a steady job. Put savings in a bank account. Earn compound interest!" Yet there are no steady jobs any more. Banks have changed, too: They're now painful to deal with, and pay negligible interest. Today's graduates will have not only several jobs but several *careers* over the course of a working lifetime. They will face globalization, cyber-warfare, climate change, the rise of artificial intelligence, and other forces against which the counsel "Get a job and a bank account" doesn't begin to suffice.

If they want to eat the future before the future eats them, they need to know what kind of creature the future is. They need skills for forming a fundamental perspective about the future, and for facing the future. (These are more important than skills for making predictions about the future, though we will also look at some of those in the

---

[1]It is popularly *thought* that Einstein and Berra said these things. We're not sure. See https://quoteinvestigator.com/2012/12/06/future-not-used/

© Springer Nature Switzerland AG 2019
F. Phillips, *What About the Future?*, Science, Technology and Innovation Studies,
https://doi.org/10.1007/978-3-030-26165-8_1

following pages.) UNESCO agrees; the UN agency has pledged to "pave the way towards establishing anticipation and future literacy as a normal component of education and training for everybody."[2]

We will now start to explore ways of conceiving of the future—the inevitable future.

1.1   Plan of This Book

The key feature of the future is that we (mostly) don't know what's going to happen. (Though we tend to be more certain of what happened in the past than we are about what's going to happen in the future, we are really not totally sure what happened in the past, either.) The ideas of future, uncertainty, and risk are tightly bound together. The next section of this chapter briefly explains why uncertainty about the past actually increases our uncertainty about the future.

Chapter 2 goes into more detail about risk and uncertainty, clarifying concepts we'll need for subsequent chapters, and highlighting another tightly bound strand, *decision making*.

Chapter 3 details the relationship between values on the one hand and concepts of the future on the other.

Now that we've established that we don't know the future, Chapter 4 explores why we might *want* to know it. The reasons span the psychological, the organizational, and other spheres of life. One of these spheres has to do with our (and our organizations') values.

Chapter 5 deals with applying knowledge of the future to personal and organizational planning. Chapter 6 goes on to highlight some of what the humanities and social sciences have to say about our perception of the future.

The future may or may not be more complicated than it used to be. It *seems* more complex now because we are inundated with more information every day than ever before. So, for all practical purposes, we can say the future is unprecedentedly complex. Luckily, we have new tools for making sense of this complexity. They are explained in Chapter 7.

Chapter 8 explores the extent to which mathematical analysis helps us see the future, and the challenge today's "big data" presents to our mathematical abilities. Chapter 9 abruptly changes direction from quantitative analysis, focusing on the role of *imagination* in seeing—and creating—our futures. Can "experts" see the future better than we ordinary folks? Much business and political prognostication depends on expert opinion, a topic examined in Chapter 10.

In at least two ways, the present dictates the future. Chapter 2 mentioned one of them: that the future does not arrive everywhere at once. "Spatial diffusion" means that some cities, states, or countries adopt new technologies and practices before

---

[2]https://webmagazine.unitn.it/evento/sociologia/57877/first-international-symposium-of-unesco-chairs-in-anticipatory-systems

others do. Thus, we can often see the future today, by traveling. For example, I first saw successful social adaptation to high-density city life in Japan in the 1970s. I first saw articulated buses and unattended gasoline pumps in Switzerland, before they appeared in the USA.

Another way the present dictates the future is when a trend becomes effectively irreversible. This is something like the "tipping point" idea made famous by Malcolm Gladwell. Chapter 11 lists a number of trends that have reached this point of no return. I've collected these as a hobby for some years. I hope you find them as fascinating as I do.

Turning what we think we know about the future into public policy is not easy. Chapter 12 delves, if shallowly, into this contentious topic.

In Chapter 13, we return to psychological ideas, portraying the future as a tension between our expectations and unfolding realities. Chapter 14 concludes the book with a discussion of the "future of the future," or how our evolving view of the future will shape actual futures—and what current events might hobble our ability to foresee.

Though some of these topics may seem—and are—quite elementary, books and courses rarely make them explicit. By turning fuzzy, implicit understandings into precise, explicit understanding, we prepare ourselves for practical forecasting and foresight projects. This book provides a "back to basics," useful philosophy of the future.

In the next section, we begin to explore what "the future" means.

1.2   A Determined Future?

A simplistic way to think of the future is as a "timeline."

There's no fuzziness in the line of Fig. 1.1. It doesn't diverge. It just keeps on keeping on.

The line shows what we might call a *deterministic* view of the future. It is a view that is common in a number of traditional religions, and in the "clockwork universe" conception of physics that was common between the times of Isaac Newton and Albert Einstein. It held that God, or the "clock" mechanism driving the universe, had predetermined everything that has happened, is happening, and will happen. Nothing we can do about it. *Que será, será.*

If we cannot predict the future, said this view, it is only because the universal clock, or the mind of God, is too big and complex for our minds to encompass.

Past   Present   Future

Fig. 1.1 Time as a simple timeline. Source: Phillips (https://webmagazine.unitn.it/evento/sociologia/57877/first-international-symposium-of-unesco-chairs-in-anticipatory-systems)

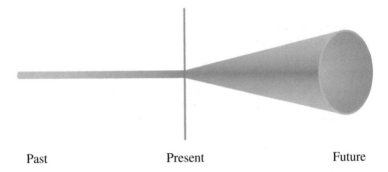

Past   Present   Future

Fig. 1.2 Time as determined past, uncertain future. Source: Phillips (https://webmagazine.unitn.it/evento/sociologia/57877/first-international-symposium-of-unesco-chairs-in-anticipatory-systems)

As you can imagine, this view sets a psychological damper on the kinds of decisions its adherents feel free to make, about their lives and about the future.

I called this view "simplistic." Yet it is, if not more advanced, at least more structured than the view of other traditional societies, which was that the gods are totally whimsical, and their whimsy is felt on Earth. In that view, earthly events are chaotic, and there would be no point in trying to make sense of anything. Earthly causes would not necessarily have effects, and thus the progression of time would be confused.

This last view was in some agricultural communities combined with a *cyclical*, or circular, concept of time: summers, winters, births, deaths, sowing, and harvesting. With no technological innovation nor contact with other cultures, time seemed only to repeat itself. There was no "progress," and so, no linear progression of time. Predicting the future, at least in broad outline, was easy. It was the past, come around again.

1.3   Undetermined Past, Fuzzy Present, and Uncertain Future

Two changes in mindset resulted in the different view of time that is shown in Fig. 1.2. In this view, the past is still determined, but the future is not. In fact, the farther future is considerably less determined than the near future.

One of the mindset changes is the idea that human action—our volition—can change the future.

The other change is the admission that there is uncertainty in the scheme of the universe: uncertainty in the actual fabric of the universe, and not just uncertainty in our minds. In other words (for believers), God is *sometimes* whimsical, or (for atheists), there is no universal clock!

Thus, in Fig. 1.2, the uncertain future is portrayed as what physicists call "a cone in space-time."[3]

---

[3]These ideas were first developed in Phillips (2008).

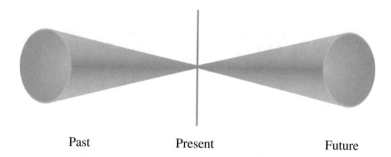

Past   Present   Future

Fig. 1.3 Past events are admitted to be uncertain. Source: Phillips (https://webmagazine.unitn.it/evento/sociologia/57877/first-international-symposium-of-unesco-chairs-in-anticipatory-systems)

Why is the past, in this picture, still a thin line? Consider that we rarely hear a historian utter the standard scientific disclaimer "at the present state of our knowledge, we tentatively believe the proposition might be true." Irreducible uncertainty in a historical study is not portrayed as a statistical residual, but rather as . . . religion! "We believe the artifact served ritual purposes." "The complex appears to have been used for worship."

In other words, the researcher simply doesn't know what function the excavated site originally served, but is covering up ignorance with the religion ploy. As a result, archeology books are written without acknowledgment of uncertainty, and an uncritical reader would conceive of history as determined. National mythos has the same effect: Differing views of the importance of racism and genocide in US history are sanitized out of school textbooks. Thus the portrayal of the past as a straight line in Fig. 1.2.

Our courts of law see it differently. Witnesses deliver conflicting testimony about a past event, and physical evidence is ambiguous. Juries deliver a consensus picture of the past, based (in US civil cases) on "preponderance of evidence," or in criminal cases, "beyond a reasonable doubt." Physicists, likewise, tell us Schrödinger's cat is only dead or alive after its box is opened; before it is opened, the cat's status was "undetermined." The legal and physical view of time looks like Fig. 1.3.

As foresight scholar Michel Godet, has remarked, "The past is as multi-faceted and uncertain as the future."

Psychology and neurology now tell us there is a time lag between an event and our perception of that event and, moreover, that perception is colored by our biases and our prior expectations. In this fourth view, time looks like Fig. 1.4.

Appropriately enough, time looks like an hourglass (laid on its side, in Fig. 1.4), widening as we go farther back or forward, and narrow—but definitely not the dimensionless point that would denote full certainty—at the "present moment."

This view is important for future gazing in business and in policy making. First, because it recognizes the difficulty of *explanation*. A butterfly may cause a hurricane, as modern complexity theory tells us, but we can't identify the butterfly by

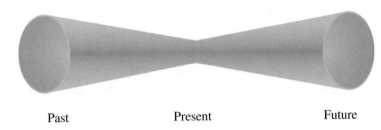

Past   Present   Future

Fig. 1.4 Psychological and neurological issues mean our perception of the present is not exact. Source: Phillips (https://webmagazine.unitn.it/evento/sociologia/57877/first-international-sympo sium-of-unesco-chairs-in-anticipatory-systems)

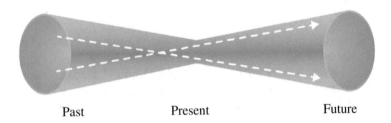

Past   Present   Future

Fig. 1.5 An event could have two different plausible pasts and two different plausible futures. Source: Phillips (https://webmagazine.unitn.it/evento/sociologia/57877/first-international-sympo sium-of-unesco-chairs-in-anticipatory-systems)

analyzing the hurricane. A current trend may arise from multiple histories, and the new view allows multiple possible histories and explanations.

Second, multiple histories (explanations) allow more latitude in extrapolating, and perhaps therefore a more realistic view of what extrapolation can and can't do. Figure 1.5 shows that, in principle, we can make two extrapolations of the same phenomenon from two different but perhaps equally plausible histories.

Look at the point where the two dotted line arrows intersect. It represents an event. We might wish to explain the event (by describing the past that led to it), or to forecast its consequences (the future that stems from it). The hourglass of time allows many straight-line paths from the event's past to its future. The dotted arrows represent two of these paths. Neither of the trajectories shown in Fig. 1.5 would be possible if time were conceived in the earlier ways.

It understates the matter to say that constructing time in this way adds to the challenge of forecasting. Yet it seems to be a realistic way of viewing time, and it may help prevent overconfidence on the part of forecasters.

It is also practical. One writer (Callick 2007) has made the chilling allegation that the government of the People's Republic of China forces academic historians to take a deterministic and prescribed view of history. Figure 1.5 shows why this policy will constrain that country's view of its future as well as of its past.

Recently developed "blockchain" technology creates records that cannot be erased or altered. By firming up our record of the past, blockchains—when they are used—will help us get a better handle on the future.

Key Takeaways

- The future is getting more complicated. Each of us needs to form a philosophy about what the future is or may be.
- Our present culture views time as moving in one direction only, with causes always preceding effects. Some historical cultures conceived time in different ways.
- The most realistic of viewing time is that neither the past, the present, nor the future is determined. Thus, there are multiple ways to explain and extrapolate the same trend.

References

Callick R (2007) The China Model, From the November/December 2007, issue of The American. http://www.american.com/archive/2007/november-december-magazine-contents/the-china-model

Phillips F (2008) Change in socio-technical systems: researching the Multis, the Biggers, and the more connecteds. Technol Forecast Soc Chang 75(5):721–734

Risk and Uncertainty

2

This too shall pass.
—Sufi poets

The future is uncertain and risky. In this chapter, and again in Chap. 5, we turn our attention to how companies, policy makers, and individuals handle risk.

Let's start by defining risk—or rather, by exploring how others might define it. Figure 2.1 shows how an investment of $11,000, less a $100 sales commission to the broker, might in a year turn into $13,000, or might turn into $9000, each possibility carrying a probability of 50%. What is the "risk"?

Some investors will say the risk is that there's a 50–50 chance of winning or losing money in the stock market. Others will define risk in another way, namely, that there is a good chance of losing $2100, that is, 11K–9K, minus the $100 sales commission. Still another will include the opportunity cost of not using the bank account, and say the risk is the chance of losing $2220, which is 11,220 minus 9000. Sophisticated investors might use the calculus of expectations to devise more elaborate ways to conceive the nature and magnitude of their risk.

> The World Economic Forum's 2018 *Global Risk Report* tells us to expect 2018 to be a potentially volatile year. About 59 percent [of WEF's survey respondents] expected to see risk increasing, and 93 percent said they expected a worsening of "political or economic confrontations/frictions between major powers" this year. (Saldinger 2018)

The first part of this quotation equates risk with volatility. It seems to mean that the measures of many political and environmental conditions will be sharply different from what they were in 2017. "Chief among the global risks is the environment and climate change," the report reports, ungrammatically and without complete logic. Perhaps it means measures of environment and climate will be volatile.

"The survey found that all five risks in the environmental category were ranked higher than average on both likelihood of happening and on impact over the next 10 years." Now we're talking! The WEF cleverly punts on whether risk is

© Springer Nature Switzerland AG 2019
F. Phillips, *What About the Future?*, Science, Technology and Innovation Studies,
https://doi.org/10.1007/978-3-030-26165-8_2

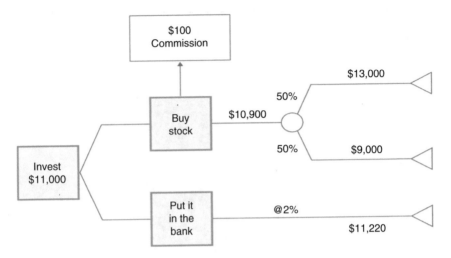

Fig. 2.1 A decision tree helps us understand risk. Image adapted from www.decision-making-solutions.com/decision-making-tree.html

"likelihood of happening" or "impact," letting readers take their choice, or even multiply the numbers together if they so choose.

There is in fact no single scientific definition of risk. This implies we should ask coworkers what they mean by "risk," before moving ahead with any project that might suffer from misunderstandings among the team.

We will follow the economists[1] in making the distinction between *risk*—the unknownness of future outcomes for which we know the probability distribution (e.g., the well-known bell-shaped "normal" curve)—and *uncertainty*, which refers to unknown outcomes with unknown probability law. It stands to reason that in the short run, much of the unknown future can be characterized as risk; in the long run, it's almost all uncertainty. This is illustrated in Fig. 2.2. The figure conforms to the convention that our ignorance of the farther future is always greater than our ignorance of the nearer future.

Figure 2.2's suggestion that uncertainty completely dominates risk after 50–75 years will be explored later in this book.

The sophisticated statistical mathematics of trend extrapolation is called "time series analysis." It is based on assumptions about probability distributions, and so is associated with risk rather than with uncertainty. The figure thus provides another reason why trend extrapolation breaks down in the long run: uncertainty takes over.

Even in the short run, we attend too closely to risk and not enough to uncertainty. Mitsui Bank traded commodities and carefully studied the risk inherent in the variation of commodity prices. The bank's fall, due to a renegade copper trader who did not report his losses, was a consequence of uncertainty, not risk.

---

[1]For example, Knight (1951).

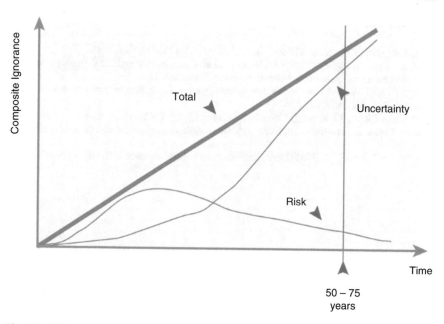

Fig. 2.2 Risk, uncertainty, and the unknownness of the future. Source: Phillips (2001)

The magnitude of risk depends not only on the size of potential markets and the absolute size of the investment needed to exploit them, but on the fraction of the firm's resources that must be devoted to the development project. Boeing "bet the ranch" on the 767 (and might yet lose it due to the faulty 737MAX), while IBM exposed itself to negligible risk in its initial development of its personal computer (McGrath et al. 1992).

We cannot always quantify risk and uncertainty as the variance of a financial cost or revenue. Psychological and social uncertainties are real (Schiffman and Kanuk 1987) (though we cannot see them!) and can affect the outcome or success rate of a business venture. A good current example is fear concerning security and privacy on the Internet.

Key Takeaways

- We tend to focus too much on risk (probability distribution known) and not enough on uncertainty (distribution unknown) when predicting the future, both in the long run and short run.
- Yet any two people are likely to define *risk* in different ways. One might say, the risk is the maximum possible loss, while another may say, the risk is the probability of a loss. A third might say it is the "expected" loss. There is no "official" definition.
- Besides financially quantifiable risk and uncertainty, we should attend to psychological and social uncertainties. These too can impose great impact on the outcome of a venture.

References

Knight FH (1951) The economic organization. Harper Torchbooks, New York

McGrath M, Anthony M, Shapiro A (1992) Product development: success through product and cycle-time excellence. Butterworth-Heinemann, Stoneham, MA

Phillips F (2001) Market-oriented technology management: innovating for profit in entrepreneurial times. Springer, Heidelberg

Saldinger A (2018, 18 January) Climate, cybersecurity top list of global threats in new report. https://www.devex.com/news/climate-cybersecurity-top-list-of-global-threats-in-new-report-91899

Schiffman LG, Kanuk LL (1987) Consumer behavior, 3rd edn. Prentice Hall, Englewood Cliffs, NJ

Values and the Future

3

The future will be better tomorrow.
—John D. Patten

Values are things the top managers believe—and want employees to believe—about the company, the workplace, the market, and about getting along with each other. From 1995 to 2004, my academic department's (the "MST" Department) strategic plan listed these values:

> We believe in these things about education, business, and society:
>
> - Better management education will increase people's and companies' opportunities for success and decrease waste and ethical lapses.
> - Fast technological and social change requires managing across boundaries, via:
>   - Interdisciplinary research
>   - Cross-sectoral alliances
>   - Teamwork across corporate departments and across companies
>   - Finding opportunities and effectively teaming and negotiating across national borders.
> - Technology drives economic development, first at the regional level, and then at national and international levels. Regions can and should take charge of their own transformation by making choices about technology-based economic development.

(continued)

© Springer Nature Switzerland AG 2019
F. Phillips, *What About the Future?*, Science, Technology and Innovation Studies,
https://doi.org/10.1007/978-3-030-26165-8_3

We believe these things about how MST should be managed:

- Alliances increase our customer base, access to resources, research quality, flexibility, and level of collegiality.
- In making and fulfilling moral contracts.
- In management by objectives, with each faculty member making an annual agreement with the department head.
- In leverage and synergies. Economies are only achieved if we talk with each other about programs we are managing; so we must also believe in communication.
- In selling and in market research. Every external contact is an opportunity to present MST's programs to a potential customer or friend of the department. Every external contact is an opportunity to ask what curriculum would best serve that contact's firm and industry. We always take these opportunities.
- In being responsive, rigorous, and relevant.
  - Responsive. Convenient classes, flexible scheduling, and excellent customer service.
  - Rigorous. Small classes with great instructors at a highly reputed research institution, producing graduates who are responsible, ethical, capable, and committed to learning.
  - Relevant. Tight focus on technology industries. Faculty with industry experience and academic credentials.

It's not always clear what "value" values have, or whether announcing them changes anything operationally. However, it does seem pretty clear that things wouldn't have gone so wrong at Enron[1] if everyone had signed onto a values statement. As Walt Disney's brother Roy declared, "It's not hard to make decisions when you know what your values are." The value statement in the box above did influence the way our department chose curriculum directions and allocated resources.

Then too, if you don't spell out your values explicitly and test your programs for compatibility with your values, customers will look beneath your programs to make guesses about what your values are—and then judge you on these supposed values. Both Microsoft and Apple, for example, originally used slogans that suggested underlying values. Microsoft's "A PC on every desk, and Microsoft programs on every PC" was totally about quantity, and not at all about quality. Apple's "Insanely great products" tagline was about quality—but essentially said, "Hey, look at me!" Neither slogan reflected caring about the customer, and the results have manifested through antitrust suits and sinking market shares.

---

[1]https://www.investopedia.com/updates/enron-scandal-summary/

Technological innovation and technology entrepreneurship—which change our world—are often motivated by high ideals. Innovation and entrepreneurship can make the economic pie bigger by increasing productivity and reducing waste and pollution. They produce wealthy individuals who then engage in philanthropic activities that benefit the public in ways governments do not. These motivations embody values. To be sure, they are open to argument from differently motivated people. Thus, they provide a good starting point for a discussion of the limitations of forecasting.

3.1   Values Are Not Negotiable

A company's overall mission, goals, and values are identified in a strategic or long-range planning process. The resulting strategy is a template for producing short-range plans. The short-range plans show how the company will accomplish objectives that contribute to progress toward the overall goals. Two facts of the planning process are pertinent for us now: Goals may be fuzzy, and are negotiable, and values are not negotiable.

Individual people may have to compromise on their values in life-or-death situations. If a business firm goes bankrupt by honorably adhering to its values, "it's only money." Firms and other organizations should stick to their values come hell or high water—both for moral reasons and because the press will clobber them if they don't.

3.2   Example: Acer Computer

A faculty member in my department, who had been COO of Taiwan's Acer Computer in the 1980s, recalled the company's initial 5-year plan. It was decided that Acer's core value was "People are basically good." (Before you dismiss that as simplistic, be aware that the question of innate goodness and evil had been continuously disputed over the course of a millennium of Chinese philosophy.) That simple decision affected everything about the way Acer treated its employees, customers, suppliers, and stockholders. At the end of the 5-year horizon, in 1990, Acer had achieved its stated goal of growing revenues to US$1 billion. As of this date, there has been no deviation from the core value "People are basically good."

3.3   Fuzzy Goals and Firm Values

US National Medal of Technology winner George Kozmetsky often drew some variant of Fig. 3.1 in his blackboard talks at the University of Texas' IC^2 Institute. The figure suggests that goals are chosen to be consistent with values; that values do not change with circumstances; and that the company makes irregular progress

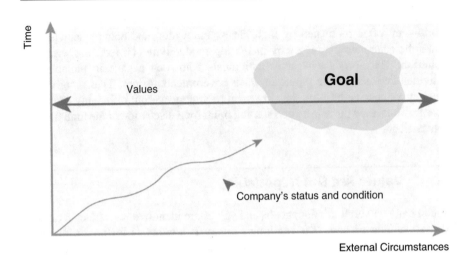

Fig. 3.1 Goals are flexible; values are not. Source: Author

toward fulfillment of the (possibly fuzzy) goal. Risk arises, of course, from variation in all the factors that make up the horizontal axis.

The question of values cuts to the heart of forecasting, for the following three reasons:

1. Technology and ideology are the dual drivers of socioeconomic change. This is the view that Kozmetsky had advanced at Texas for more than 30 years, and to which more and more communities and governments are being converted.
2. One person's favorable outcome may be very unfavorable to another. If an outcome threatens a person's deeply held values or beliefs (abortion or irradiated food are possible examples), the most adamant resistance ensues (Rogers 1983).
3. One man's unshakeable faith in an outcome (at the craps table, in the hereafter, or wherever) may be a source of great uncertainty or anxiety to the next man.

We may agree that the American business planning horizon is too short and that "net present value" is not always a sensible planning criterion. Yet in the social context from which we will draw many of our next generation of employees, there is not even this rudimentary apprehension of the future.

Some years back, a radio interviewer asked a teenaged Los Angeles gang member what he expected to be doing in ten years. The young man replied, "I expect to be dead in ten years." How can you blame such a person for not having a sense of the difference between capital and yield, between spending and investing? What reason does he have to invest in the future? Do those of us with greater education and privilege share basic values and transmit them to posterity? Surely those values do not include allowing this young man's life to end in the way he expects and which reflects so poorly on our society. How have we allowed this to come to pass, and what does it mean for our future?

Let's look at some very small and very large examples of how social values and institutions interact with technology. Suppose, first, that there was a micromachine that could be cheaply introduced into rivers in great quantities. The machine would swim to the ocean, collect proteins, and convert them to a form that is not only edible but also delicious when gently heated. Not only that, but the machine would, after a year, swim back upriver to its point of origin! Wouldn't investors be beating down the door to have a piece of such an enterprise? There is such a machine, of course, but there is currently no way to invest in salmon runs except through the tax-supported hatcheries and so on. Other business interests, shipping and cattle grazing, for example, better fit our society's dominant model of private enterprise and private property, and so have been encroaching on salmon habitat with alarming success.

People want salmon and pay high prices for it. Here is a case of antiquated institutional structures interfering with nature's hugely efficient production of a valued food resource. My prediction is that institutions will change, but not without a lot of shouting and litigation, and in ways not currently predictable. For example, if I were to patent the world's first blue-eyed salmon and build a private hatchery for them, then any blue-eyed salmon caught in public waterways would be mine—or would it? The question would certainly be tested in the courts. It illustrates a contentious issue that will become still more contentious before (one hopes) it is solved: The private sector wants small government and privatization of public assets, yet the private sector is crippled by a short-term outlook forced on them by global capital mobility.[2] It is only governmental bodies, becoming smaller and weaker under currently dominant political philosophies, that think about preserving public assets for posterity.

3.4   Long Time Scales for Macroengineering Projects

At the other end of the size scale, there is a clear relationship between long-range planning and macroengineering. The pyramids, Stonehenge, the Suez and Panama canals, the space programs, and even some of today's large buildings and oil tankers take a long time to build and require faith in, or good forecasts of, political stability, commercial demand, availability of materials and qualified labor, and cost of capital.

The issue of values arises when we consider how some of these macro-projects were paid for. Many were motivated and financed directly or indirectly from war efforts. Others are straightforward monuments to despotic regimes. Still others exploited the religious faith of populations or exploited laborers who had no alternatives.

---

[2] And by CEOs, wanting to retire on a high note, who boost stock price by cutting "costs," weakening the organization's ability to thrive in the future. A 2018 study (www.thelowdownblog.com/2018/03/study-when-ceo-equity-is-about-to-vest.html) showed that executives also do this just before their stock options are about to vest.

The American industrial barons of the nineteenth century built a splendid network of railroads, on the backs of underpaid workers, then used the profits to donate libraries and other civic works to the American public. Hindsight tells us that these libraries were instrumental to the success of the American experiment and to our individual lives. There is debate about whether the donors' purpose was to elevate the level of public discourse or to "tame" the immigrant workforce (Tisdale 1997).

It is hard to come to an ethical judgment about the origin of these public works. One possible consideration is that public works should be public decisions. If Congress squeezes my take-home pay to build an unneeded airport, I can vote the bastards out. If my employer squeezes my wages to finance his charitable contributions, that's taxation without representation, exactly the reason a revolution was once fought in my country.

What values are represented by readers of this book? The publisher's market research suggests "sustainability," "health," and "space travel" are prominent among them. Are we capable today of pulling together for the long-term efforts these goals will require? Macroengineering projects like railroads, the Internet, or the space programs can be beneficial, or even vital to human survival. Yet building them can require outright tyranny, or alternatively a less reprehensible but still unpleasant suppression of diversity behind a false front of unity, such as was the case in the American society of the 1950s. It may not be very profound to observe that overt pluralism can be the enemy of macroengineering, but it is hard to say it without qualms of conscience.

3.5   Nation States Becoming Less Important

Having mentioned the weakening of national governments, we should entertain some ideas about their long-term future. Individual nation states do seem to be on more shaky ground, and new institutions of a number of kinds are taking on increasing importance. Since the 1970s, there has been tension between the practices of modern multinational firms and the sovereignty of nations. At the least, the firm's internal but cross-border transfers of goods and funds have made the compilation of economic statistics, and hence forecasting, more problem-laden. At most, it has become impossible to say that a company is "an American company," or a "French company," etc.

Novelist Poul Anderson (1996) wrote,

> In practice, though private organizations exist on sufferance of the state, the real dictator is always the man who controls the armed forces and the police. At most, large corporations may be junior partners of government—very junior—and this is possible in just a few countries. Other outfits, such as unions, could as well fill the role, and churches have sometimes been coequal or senior.

Yet David Wood (1997) of Newhouse News Service maintained, "In the twilight of the 20th century, national governments are losing their monopoly on armed violence." In a *Harper's* article, Elizabeth Rubin (1997) describes a South African mercenary corporation that is fighting wars on behalf of a number of African governments. Anderson himself noted that wars between city-states in Renaissance Italy were fought by mercenaries, and that lacking a cohesive ideology or loyal forces, Italy became easy pickings for foreign powers. Thus, the loss of citizen-based martial power feeds back to further deterioration of the nation state. We might therefore note with alarm the rise of "private contractors" taking on military missions under contract with the US armed forces.

Ethnic tensions tore Yugoslavia apart and may do the same to Spain and Canada. Yet when separation is finalized, there is no ready way for the new ethnic enclave to organize—and to be represented in bodies such as the United Nations—except as a nation state. So, there is a countervailing force, creating new states with perhaps stronger ideological unity than before.

Another countervailing force might seem to be the nationalist politics that have gathered momentum in the USA and several European and Asian countries at the time of this book's publication. However, this "nationalism" seems to be a thin cover for corporatism, through renewed trade protectionism. Thus, by benefiting corporations without strengthening national culture, this kind of nationalism does not strengthen the nation state, but rather hands to the private sector what had been the influence of government and the people.

3.6 NGOs

Although technology entrepreneurs live in communities like everyone else, their arena of business and competition is global. They feel this split in their lives rather acutely, and so embrace new kinds of organizations, like the Austin Technology Council or the Technology Association of Oregon with which I have been involved, which bring together global issues and community concerns and clarify their mutual relevance.

New institutions—and by this I mean not just abstracts like "liberal democracy," but specific groupings of people like the Technology Council—can also be publicly supported middlepersons that spring into existence where gaps in training or knowledge need to be filled. A third purpose of these new organizations is to increase the intensity and quality of communication among members, in the context of rapid change and info-glut. The organizations may be nonprofit or for profit, but are "entrepreneurs" in the literal French meaning of the term. They are industry associations, professional societies, civic organizations, user groups, or hybrids, but typically engage members from more than one sector (university, government, industry, press) to rectify a situation in which the traditional activities of one sector have failed to meet new imperatives.

This is a long way of saying that non-governmental organizations (NGOs, usually known in America as nonprofit organizations), along with international organizations like the World Economic Forum, may play pivotal roles in our future. The press is weakening (due to corporate acquisitions and competition from bloggers), and fewer young people belong to religious organizations. Public universities are struggling for funding, and private universities have turned into "hedge funds with universities attached for tax purposes" (Taylor 2016). This leaves NGOs, local governments, and the remnants of national governments as the principal counterbalances to runaway corporations.[3]

Sadly for business forecasters, we cannot now foresee which of these new institutions will be seed crystals for massive new economic activity and which will be short-lived and forgotten. The remedy will be to declare that "institutional studies" be no longer a cuss word in the economics profession. (This has begun to happen since the first drafts of this chapter were written.) Published studies of these new institutions will give us all a better idea of what is happening and what is going to.

Key Takeaways

- Values are not negotiable. Values guide the actions an individual or a company takes toward reaching goals and objectives. Goals and objectives, unlike values, can be adjusted.
- In the future, the role of nation states may be dominated by new institutions such as non-government organizations, giant private corporations, or international organizations. This suggests that more "institutional studies" about these new (or revived) types of organizations are needed.
- The massive infrastructure projects that are sorely needed in the USA and elsewhere require long-lived supporting institutions that are consistent in their visions. They also require debate about what is a private good and what is a public good.

References

Anderson P (1996) All one universe, New York, St. Martin's Press

Rogers E (1983) Diffusion of innovation, 3rd edn. The Free Press, New York

Rubin E (1997, February) An army of one's own. Harper's Magazine: 44-55.

Taylor A (2016, March 8) Universities are becoming billion-dollar hedge funds with schools attached. The Nation

Tisdale S (1997, March) Silence please: the public library as entertainment center. Harper's Magazine: 65–74.

Wood D (1997, February 23) Rogue warriors. The Oregonian. Newhouse News Service, p E1

---

[3]A little corporatism is good for us. You may not agree that a lot of corporatism is bad. However, I hope you will agree that each sector of society, including the corporate sector, needs to be balanced by equally vital other sectors—whether these be government, NGO, press, academe, or church.

Why Do We Want to Know the Future?

4

Life is what happens while you are busy making other plans.
—John Lennon

In this chapter, we look at how we can prepare our minds and our organizations to regard the future in an effective way.

We are motivated to think about the future—and often, worry about it—for many reasons. We want to gain a material advantage by knowing what will happen. We want to avoid the wasted effort of preparing for an anticipated event that will not happen. We want to resolve our anxiety about the outcome of a situation.

In other words, we have *goals*: to have enough to eat tomorrow, to win in a gamble or investment, to leave our umbrella home if it will not rain, or to make it to the next filling station when the gauge is on empty.

People who do not have goals don't have to think about the future!

A demon of uncertainty stands between us and our goals. Uncertainty about what may or may not happen in the future affects our chances of realizing the goals.

Flowers can't bloom without soil, and goals can't spring forth without *vision*. A vision explains why you have adopted your goal. "To boldly go where none have gone before." "To increase our customers' enjoyment of life by creating affordable entertainment appliances using micro-widgets." The vision is a human or social context for the goal.

A goal can be executed effectively even by a group of people with vastly differing motivations. This can mean, though, that a strong-arm management style has substituted for a shared vision. A well-articulated and effectively shared vision is a fine motivator and a great tool for strategic alignment in the organization.

The vision explains and justifies the goal, enabling one and all to focus more clearly on the future.

© Springer Nature Switzerland AG 2019
F. Phillips, *What About the Future?*, Science, Technology and Innovation Studies,
https://doi.org/10.1007/978-3-030-26165-8_4

4.1   Right, or Ready?

In another sense, there are only two reasons to anticipate the future: to be right or to be ready.

Scholars, investment analysts, and intelligence agencies spend huge efforts trying to make their forecasts right—that is, to make them more precise and accurate, to make predictions that, some time later, come to closely match the actual outcome. It turns out, though, that there are very few instances in which being right is important!

The person who makes the most accurate forecast earns the right to do a happy dance, pump a fist in the air, and perhaps be introduced as a new sage, on a television talk show. And naturally, the person who puts his last chip down on number 18 as the roulette wheel starts to spin hopes to be right.

Even then, though, it is as important to be ready as to be right. Is the lucky forecaster ready to be a media darling? Ready to live with continued or renewed obscurity if trumped by another forecaster? Is the roulette player ready to be richer? Ready to go home busted, if the ball falls on 23?

Did you just say to yourself, "Sure, everybody's ready to be richer"? Consider the many lottery winners who have been completely unprepared for taxes, investment scammers, demands from relatives, and the sudden reappearance of long-lost "friends."

The purpose of forecasting is not to be right, but to be ready.

A chess master uses her knowledge of her opponent's past games—and of the opponent's psychology—to predict the opponent's next move. She is *more ready* to respond to that most likely opposing move, in the sense that her strategy for responding to it is more completely formed. Of course, she is also prepared for the opponent's less likely moves. That is what makes her a master.

4.2   Predicting for Advantage

At lunch with an Asian friend, I joked that we could go into business together when he returned to Japan. From the other side of the International Date Line, he could send me tomorrow's stock prices, and we'd split the profits. My friend found this idea amusing, or politely pretended to, but he understood the proposition was absurd.

Absurd, first, because although Wednesday morning in Japan is Tuesday night in America, it is really the "same instant" in what we might call universal time. There is no stock movement in this instant. Understanding this, brokers in Hawaii wake up at 3 a.m. Honolulu time, in order to be online when the mainland US markets open.

And absurd in a second way because if my friend and I could do it, everyone could do it, wiping out any advantage from information asymmetry. This points to another reason people want to predict the future: they want to be the *only* one to possess the prediction. Thus, discreet visits to Madame Zelda, who with her crystal ball "knows all and sees all."

"Will my marriage be happy?" the seeker asks. "Will a new job improve my situation? Should I invest in cryptocurrency?"

Only those smart enough to visit Mme. Z, thinks the deluded seeker, can benefit from knowing what will happen tomorrow. On occasion, Zelda hits it right on the button. (Indeed, in any collection of sages, logic tells us one of them is bound to get it closest to "right.") The seeker then exclaims to his friends, "Mme. Zelda is amazing! She has true powers of foresight!" Zelda posts this encomium on her Facebook page. Zelda's remarkable success is so memorable that it crowds out her customers' recollections of the dozens of times Mme. Z got it wrong.

This points to the connection between superstition and the popular misunderstanding of probabilities, and also to the phenomenon called selective memory. In later chapters, we will revisit questions of psychology and probability. Our focus for the remainder of this chapter concerns being *ready,* and the idea of *goals.*

4.3 Flexibility

Readiness means being flexible enough to respond constructively no matter what kind of situation we are facing. This is true for our individual lives, for our business organizations, as well as for our military.

Every culture has a fable that illustrates this. In many, the oak tree is compared to the willow. See Fig. 4.1. The willow can either stand upright or bend. The oak can only stand stiffly upright. Both thrive in ordinary conditions, but in a very high wind the rigid oak may fall. The willow has shown the flexibility needed to survive differing conditions.

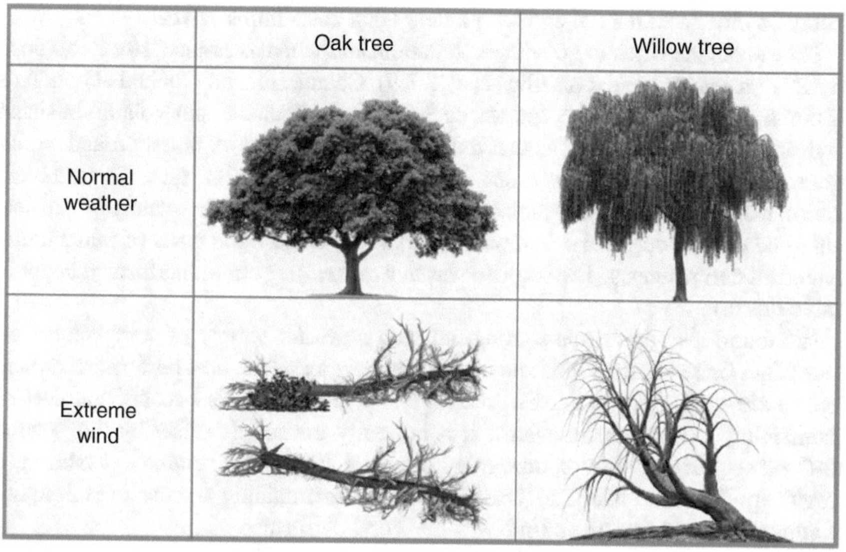

	Oak tree	Willow tree
Normal weather		
Extreme wind		

Fig. 4.1 The flexible willow survives the storm; the rigid oak does not. Source: Author

It's a nice fable, well beloved by karate instructors. Yet the willow grows near waterways—it needs water in its limbs in order to flex—and so in drought conditions it might be the oak and not the willow that continues to thrive. This bit of nitpicking (I don't want to anger any karate experts!) is meant only to stress that we should carefully draw the range of conditions under which we expect an organization or a personality to be flexible. And then the organization should hold a few extra competences in reserve, in case they serve to save the day when totally unexpected conditions come along.

Strategy scholars like to talk about exploitation behavior (strengthening an existing set of organizational competences—this is the rigidity of the oak) and exploration behavior. Exploration means developing new competences which will sustain the organization as new circumstances arise. This is the suppleness of the willow. The organization's challenge is to balance resources directed to exploration versus exploitation, when total resources are limited.

It costs money (or in the case of the willow, metabolic energy) to maintain competences, procedures, and people that can respond to rare conditions, or conditions that have not arisen for many years.

Sometimes, it costs more money not to! NASA's Saturn rocket engineers are retired now and have taken their knowledge of high-orbit launchers with them as NASA moved to the low-orbit space shuttle program. Now, as talk of moon and Mars missions resumes, NASA must reconstruct the lost knowledge from scratch. An "unneeded" competence that was discarded must be regained, at considerable expense.

This is the same reasoning that argues for maintaining species diversity. If a plant or animal contains a cure for cancer, for example, and if it goes extinct before we find it, then we may or may not discover that cure the hard way—in the laboratory, without a clue from nature, and with a delay that costs human lives.

How to decide when to go with exploitation, and when to emphasize exploration? In a 2017 research project (Phillips et al. 2019), Chang, Su, and I looked at hundreds of thousands of simulated corporate decisions in both stable and volatile business environments. We found signs that *any* strategy combining exploitation and exploration produced better outcomes in stable environments than in turbulent environments. (Of course, stable environments are easier for managers to deal with, and can produce sufficient profit margins to forgive the costs of maintaining redundant competences. The surprise was that *any strategy at all* performed better in stable times.)

We found also that firms starting out with a greater variety of competences do better than those starting with one or few competences. The best performers among these were firms that rarely discarded a competence. This is because the cost of maintaining a competence, even if it is currently unused, is offset by the "search cost" of acquiring a new competence, or (as in the NASA example) restoring a previously discarded one. The "lean" strategy of maintaining an absolute minimum of competences at any given time was the worst performer.

Simulations are unlikely to give definitive answers to the efficiency vs. flexibility question. This simulation's value was to draw attention to the balance of costs of maintaining competences (salaries, document management, machinery), discarding them (severance pay, scrapping machinery, etc.), and acquiring competences (hiring costs, M&A costs, and capital investments). The simulation showed how this balance affects the success of various competence building strategies—that is, strategies for being *ready* to meet varied business conditions, in order to meet the *goal* of continued satisfactory financial returns.

4.4   Speed Substitutes for Foresight: "Make My Day"

If you're fast enough, you don't need to predict the future. An example from new product development will explain this.

As product development projects proceed, it gets more expensive to make changes in the design. Changes are easy and cheap early on in the project, and difficult and pricey as the product launch date approaches. Especially in tightly coupled designs (in which changing one design element necessitates changing several others), ECOs (engineering change orders) are dreaded.

It's not uncommon to see, pinned to an engineer's cubicle, a printout of actor Clint Eastwood brandishing his long-barreled Colt, with a scrawled caption: "Go ahead, make one more change."

In the 1980s, General Motors' total design cycle for a new model car was 10 years. Honda's was 4 years. Two automotive innovations, fuel injection and airbags, became cost-effective and available in quantity near the middle of the decade. The 1990 Buick's development cycle started in 1980 and had already been under way for 5 years. Due to the fact that fuel injectors and airbags had not been foreseen, and due to the cost of mid-project ECOs, it would have been difficult or impossible to incorporate these two innovations in GM's 1990 Buick.

It was easy to incorporate them in the 1990 Honda because its development cycle began in 1986. Come 1990, this made the purchase choice—Buick, or Honda?—easy for a consumer who was safety-conscious, technology-conscious, or convenience-conscious. Indeed, it was in the 1980s that Honda and other Japanese makers took big bites out of the market share of GM and other American manufacturers who had not yet seen the value of compressing development cycles.

Compressing the cycle allows you to deliver more up-to-date technology to the customer. It also means *you don't have to predict the future*. Having taken this lesson to heart, Honda and GM have both, in 2017, got their development cycles down to 2 years. Speedy.

4.5   Right, but Not Right Enough to Be Useful

The future's broad outlines might be obvious, but the devil is in the details—details that can mean life or death, and fortunes won or lost. In other words, predictions can be right, but not right enough to do us any good.

We know this. It's the basis of the blues song about the old fellow who won't give up drinking. "Doc says it'll kill me," he growls, "but he won't say when."

In 2016, I firmly believed Hillary Clinton would win the US presidential election. Thus upon reviewing material for this book, I was astonished to see that I had written, in 2011, "Revolution at home is still a sleeper, but a charismatic leader with a clear vision, intellectual horsepower, and a good media machine could change that—either in the direction of greater democracy or increased corporatism." (Phillips 2011) This (the second possibility) is just what happened in 2016, although the intellectual horsepower was in the head of Steve Bannon, not that of Donald Trump. Evidently in 2011 I did not think it would happen so soon. Not to mention that I forgot that I'd predicted it...

Madame Zelda predicts you will find love. In the end, you find it—from a pet rabbit. Forecasting can disappoint. *Backcasting*, or controlled forecasting, can be a constructive alternative. We explore it in the next section.

4.6   Controlled Forecasting

In the late 1970s, David Learner and I put forth the idea of controlled forecasting.[1] The idea couples prediction with goals. Instead of a purely *predictive* statement ("The wind will blow westerly at 30 knots") or a purely *prescriptive* statement ("Head 15° NNW"), we combine these with a *goal* statement ("I want to go to Corfu") into a *normative* statement, "If you want to go to Corfu, and if the wind is westerly at 30 knots, a heading of 15° NNW will get you there in 20 h."

Put yourself in the shoes of a consultant reporting to a client: Your pure *prediction* of wind direction/velocity may not be worth offering, if the client will not use it. At least, it needs to be put together with other information in order to be useful. The client will bridle at the *prescription* you offer. "Who are you to tell us what to do," he will shout. "You don't know our situation. I am the management decision maker here!"

The *normative* forecast is most acceptable, as well as most useful, to the client or advisee.

Controlled forecasting caught on in the academic literature under a different moniker, *backcasting*. Backcasting has since come to mean a process of "working backward," first identifying a desired future state of affairs, then deciding what current and near-future actions need to be taken in order to realize that future. The resulting action plan is still subject to risk and uncertainty!

Readers of the above paragraph will immediately realize that speaking of "desired future" as opposed to "predicted future" implies a completely different mindset. It is a powerful one, much embraced by technology entrepreneurs and corporate planners.

---

[1]With advice from operations research pioneer W.W. Cooper. See Learner and Phillips (1979).

Remember, though, what automobile pioneer Henry Ford said: "If I gave people what they wanted, I would have invented a faster horse." This applies not only to your customers, but to *you* as well. Focusing on what you want can doom you to incrementalism. Diving into open-minded exploration, on the other hand, might yield for you the unexpected breakthrough that will enrich the world—and you.

Key Takeaways

- We want to know about the future so that we can better prepare, reduce anxiety, or gain advantage over others.
- Maintaining a wide range of competences allows us to be ready to cope with different business conditions, still fulfilling our (profit or other) goals.
- Forecasting may generate right answers, but not enough details to be useful.
- Backcasting or controlled forecasting, which combines predicting with goals, is a powerful way to lead us to a "desired future" rather than a "predicted future." Backcasting, however, might bar us from unexpected breakthrough.

References

Learner DB, Phillips F (1979) Information-theoretic models for controlled forecasting in marketing. In: Montgomery DB, Wittink DR (eds) Market measurement and analysis: Proceedings of the ORSA/TIMS marketing science conference. Marketing Science Institute

Phillips F (2011) The state of technological and social change: impressions. Technol Forecast Soc Chang 78(6):1072–1078. https://doi.org/10.1016/j.techfore.2011.03.020

Phillips F, Chang J, Su Y-S (2019) When do efficiency and flexibility determine firm performance? A simulation study. J Innov Knowl 4:88–96

Prediction and Planning

5

Well, you give me too much credit for foresight and planning.
I haven't got a clue what the hell I'm doing.
—Novelist Robert B. Parker

Circumstances, Hal Linstone often noted, change faster than our assumptions and values. Questions of values, resources, institutions, complexity, and scientific/technical progress all have to be considered if we are to make forecasts in a rapidly changing world. How can we consider all these thing so that we may better forecast the future—especially a far future—and make useful plans for our organizations?

We start to answer this question by looking at how forecasting, risk, and decision making are related. We then examine how business planning has been customarily done. The chapter ends with a look at the question of very long-range forecasting.

5.1 Forecasting and Decision Making

To do this, we momentarily back away from long-range forecasting, in order to use an example with nice, clean probabilities. Remember, in short-range forecasting we are more likely to encounter *risk*. That is, we can, even roughly, assign probabilities to future events. This is different from the *uncertainty* more prevalent in long-range forecasts, where we generally have no idea what the probabilities are.

Here again (Fig. 5.1) is the "decision tree" we saw in Chap. 2.

Our *decision* problem is to invest eleven thousand dollars in a savings account for a year, or to buy a stock. Two percent interest on the bank account is certain. The securities we have in mind may rise or fall in value. Somehow (the textbooks always gloss over this bit) the decision maker concludes that $10,900 worth of stock might

---

This chapter is an update of an address presented at the World Future Society, San Francisco, July, 1997, with added material adapted from Haynes et al. (1996).

© Springer Nature Switzerland AG 2019
F. Phillips, *What About the Future?*, Science, Technology and Innovation Studies,
https://doi.org/10.1007/978-3-030-26165-8_5

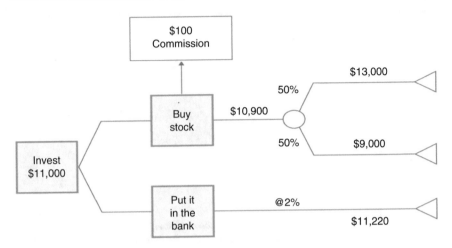

Fig. 5.1 A decision tree illustrates the connection of forecasting, risk, and decision making. (Image adapted from www.decision-making-solutions.com/decision-making-tree.html)

rise to $13K in a year, or might fall to $9K in that time, each possibility carrying a probability of 50%.

Unraveling the tree to find a $10.9K expected value for the stock investment is called *decision analysis*. It should be clear that the $13K, the $9K, and the 250% probabilities are *forecasts*. Nothing more need be said: forecasting and decision making under risk are inextricably bound.

And almost all decisions for the short term are "under risk." There's a tiny but nonzero chance the bank will go belly-up before the end of the year, leaving you with $0. Or, on your way to the bank with the $10K, a robber threatens to shoot you if you do not hand over your wallet. Skilled in intimidation, the robber talks so as to make you believe there is no risk—that you must decide either to lose your cash (with certainty) or lose your life (with equal certainty). But there *is* risk, to both of you: his gun might misfire. Police might arrive before the transaction is complete. You might be able to disarm the robber. The robber might shoot you anyway, after you disgorge your wallet. Each of these events has a probability.

In common language, it is said that the investment decision diagrammed above offers "two choices": bank it or invest it. However, the rest of this book will go down more smoothly if we are more precise in our terms. There is one *choice*, with two *alternatives*. If there were three main branches on the left (stock market, bank, or stuff the money in a mattress), there would still be one choice—what to do with the money—but three alternatives to consider in making the choice (the decision). "Choice" and "decision" have the same meaning.

5.2   Risk and Forecasting

None of the definitions of *risk* in Chap. 1 mention "upside risk." The definitions all imply that risk has to do with the downsides of future events. Yet risk is not a bad thing per se.

The businessperson wants to avoid or minimize the downside consequences of unfavorable realizations of risky situations. This is done in five ways: better information for decision making, optimal decisions, reducing the cost of wrong decisions, reducing risk directly, and pooling risk.

Better Information for Decision Making   Risk is reduced when management has accurate and timely information about the preferences and activities of customers, competitors, and sources of new technologies. This is why companies engage in market research, competitive intelligence, and technology scanning.

Optimal Decisions   Good internal and environmental information is leveraged for risk reduction by using the quantitative techniques of management science and decision analysis to make provably better decisions. An example is the use of conjoint analysis to predict consumer response to the attributes of a new product.[1]

Reducing the Cost of Erroneous Decisions   Too large a production run of electronic components? Be prepared to "dump" them, below cost, in foreign markets. A competitor has beaten you to market with the first offering in a new consumer entertainment device? Your ad agency must position your offering in a way that differentiates it from your competitor's.

Reducing Risk Directly   If left to their own efforts, the failure rate of entrepreneurs will continue at its high historical level. Companies interested in cultivating a strong base of small suppliers should support incubators and other new institutions for reducing the failure rate of start-up firms in relevant industries. Companies can participate in standards-setting bodies in order to reduce the risk of offering noncompatible products, and in futures markets to smooth the prices paid for commodities.

Pooling Risk   We hedge against extraordinary downside outcomes by devising insurance pools, building and loan societies, federal emergency services, and so on.

Some years back Nations Bank and other firms engaged in attempts to aggregate project risks into total enterprise risk. Can it work? In the financial industry, managers are not incentivized to minimize risk, but are rewarded for taking not significantly more risk than their peers. (The role of our legal and regulatory systems in controlling risk behavior in this and other industries is very relevant but beyond the intended scope of this book.) In 2019, a Google search reveals many consultant

---

[1]See, e.g., http://www.dobney.com/Conjoint/Conjoint_analysis.htm

white papers on total enterprise risk, but few academic articles. This suggests the idea did not carry much rigor.

Can the five ways of dealing with risk (better information for decision making, optimal decisions, reducing the cost of erroneous decisions, reducing risk directly, and pooling risk) be extended to conditions of uncertainty? It appears that all of them can, except for "optimal decisions," which generally rely on probabilities (When a big government project fails, all taxpayers feel the brunt. This is a clumsy way of pooling uncertainty!).

There is, as you might guess, a connection between risk and competence. One large construction firm considered itself extremely good at what it does. Accustomed to doing business on a cost-plus basis (which leaves the customer with the risk of cost overruns), the CEO decided to offer fixed-price contracts in the future. This internalized the risk. The CEO believed fixed-price contracts would attract new customers and raise the firm's market share, thus making up for possible financial losses on a few jobs.

In this case, the board of directors terminated the CEO. Business gospel holds that the function of the firm is to internalize profits and externalize costs and risks. Our poor CEO had made a rational decision, one that would probably bring benefit to the firm, but that flew in the face of the board's traditional beliefs.

The story has two morals: first, that values drive the future, and second, that changes in business models will be a prominent, if risky, feature of the future.

5.3   How Do Businesses and Governments Plan for the Future?

This book is not a how-to manual of foresight techniques. An overview of them, however, should help you come to a constructive perspective when it is time to find a manual and start your formal future-gazing exercise. Our ways of looking at the future include the following.

5.3.1   Trend Extrapolation

Extrapolation amounts to saying, "The future will be like the present, but even more so." We take the trend line of the recent past and extend it a reasonable distance into the future. It sounds simple and it (usually) is.

The trick is in recognizing the limits to the extrapolation. Moore's Law (more about Moore and his law later in this book) served the semiconductor industry and its customers well for 50 years but is now reaching its physical limits. In the aircraft industry, every new generation of plane is faster, but there will not be an airplane that exceeds the escape velocity of the Earth. (If there were, we would call it a spaceship, not an aircraft.)

So beware of gurus who talk about "exponential growth" of investments, or exponential growth of artificial intelligence. Exponential or "hockey stick" growth can proceed only so far before it levels off. Real-world growth shows the S-shaped curves discussed in a later chapter of this book.

Trend extrapolation is called a *convergent* future-gazing technique, because it results in a single number for (or single picture of) a future. Trend extrapolation assumes current trends will continue. It maps a single future, perhaps specifying a range of uncertainty.

5.3.2 Delphi

Delphi forecasting, named after the Greek oracle, is also a convergent technique. Less mathematical and more discussion-based than trend extrapolation, Delphi taps the opinions of experts. In a simple Delphi, the opinions are pooled, shared, and discussed. Procedures are in place to prevent the more assertive or charismatic experts from dominating the discussion. Discussion is followed by a second round of opinion-gathering, in which the experts may modify their earlier answers, based on their reactions to the discussion.

Delphi is a group decision technique. It draws a "most likely" future. Delphi discussions may address simple trends, but may look at discontinuities in trends, and bring in a wider range of considerations than is possible in mathematical extrapolation.

5.3.3 Scenarios

Scenario methods are formal storytelling. "Change drivers" are identified, and stories are constructed around the possibilities that the change drivers will move in desirable, neutral, or unhappy directions. A typical scenario-building exercise will result in four stories describing quite different futures. For this reason, scenario building is called a *divergent* future-gazing method, as it creates multiple futures. Scenarios map several plausible (usually divergent) futures that are not intended to be predictions.

The combination of structured technique and traditional storytelling is a powerful method of revealing possible futures, never failing to capture the attention and passion of participants. When published, they also effectively motivate nonparticipating constituents.

Divergent techniques don't result in *a prediction,* so what use are they? Remember, the purpose of forecasting is not to be right, but to be ready. Scenarios are aimed at fostering flexible planning and response. A good set of scenarios encourages the organization to become sufficiently flexible and resilient to meet any of the (usually four) scenarios, or to meet a future that turns out to fall in between some of the scenarios. And perhaps sufficiently flexible to meet a future that does not resemble any of the scenarios!

Hal Linstone pointed out that "the military are trained to work through many scenarios, not so that they will pick 'the most likely' (as business usually does) but to make them able to react well to unanticipated ones." Robert Flood, editor of *Systemic Practice & Action Research,* adds that this is also the major thrust of the learning organizations movement.

5.3.4   Foresight

Foresight exercises (popular worldwide,[2] and in France known as *prospectives*) take the view that there are many possible futures. A foresight exercise looks at both possible and desirable futures and is usually linked to a (national policy) project aimed at bringing about one or more of the desirable futures.

5.3.5   Long-Range Planning

Fashionable in the 1950s and 1960s, when long-term stability of the business environment was assumed, long-range planning fell out of fashion in the early 1970s, after the oil shock and the end of the Bretton Woods agreement. It was replaced by strategic planning (see below), which usually took a shorter-term horizon.

Long-range planning may regain popularity in Asia, due to the Asian "Great Moderation." This phenomenon, always subject to the ups and downs of Chinese economic prospects, seems to portend a long period of steady economic growth in Asia.[3]

5.3.6   Strategic Planning

A strategy specifies an organization's mission, vision, and (sometimes) values. It lays out a number of goals that support the mission, and at a finer level of specificity, a number of objectives that support the goals. Drilling down even deeper, a strategic document authorizes programs that support the objectives.

It analyzes the firm's strengths, weaknesses, opportunities, and threats ("SWOT"). This involves looking at the internal organization, and the external business environment, including the firm's competitors. It includes a "gap analysis," detailing the rivers the company needs to cross in order to move from its present condition to its desired state.

A strategy is future oriented in that it must draw the futures of the company's capabilities, the competitors' capabilities, and the business, market, technological, and political environments. These futures are described on the basis of the informed opinions of senior management and its consultants.

Depending on the academic strategic theory that senior management adheres to, future-gazing emphasis might be on internal capabilities, or on the company's stance vis-à-vis the competition. Either way, strategy formation is getting harder, due to faster technological change, technological "convergence," and the resulting blurring of industry boundaries.

---

[2]An excellent source is Miles et al. (2016).

[3]https://www.economist.com/finance-and-economics/2012/11/10/asias-great-moderation

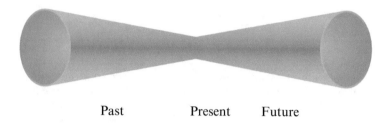

Past   Present   Future

Fig. 5.2 The hourglass view of time suggests "bounded futures" forecasting. Source: Phillips, https://webmagazine.unitn.it/evento/sociologia/57877/first-international-symposium-of-unesco-chairs-in-anticipatory-systems

Strategic planning aims to maintain a favored organizational trajectory, in an uncertain future. It has largely replaced long-range planning, at least in Western businesses.

5.3.7 Entrepreneurial Evangelism

When wooing John Sculley to Apple from Pepsico, Steve Jobs asked, "John, do you want to change the world, or do you want to peddle soda water for the rest of your life?" Jobs did not ask whether Sculley wanted to forecast the future, but whether he wanted to change it (Sculley 1987).

This is how today's "entreprenerds" deal with the risky future: First, identify a new, basic technology that will spawn whole new classes of products. Second, know more about that technology than anyone else. Third, inflict the technology on the world in the form of an audaciously marketed product. Fourth, watch for follow-on, spin-off, and tie-in opportunities—and take them.

Entrepreneurial evangelism aims to create an envisioned, favorable future—and do so quickly— usually involving a certain technological product.

5.3.8 Bounded Futures

Many thinkers, including Peter Drucker (and me!), have noted that it is easier to foresee what will *not* happen than to know what will happen. The bounded futures approach first identifies desired futures that lie outside the cone on the right-hand side of Fig. 5.2.

It then investigates any possible fuzziness in the surface of the cone. Is there any chance, it asks, that the impossible will become possible, making the cone "wider"? Meanwhile, the things currently believed to be impossible provide focus for decision making, as shown in Table 5.1.

Table 5.1 Illustrating the bounded futures approach

Example constraint	Example creative response
Traditional venture capital will *not* drive entrepreneurial growth	Develop government, corporate, angel, and crowdsourced funding for start-ups
Economies of scale will *not* drive diffusion of innovation through worldwide distribution of standard tangible products	Online networks for peer-sharing of local problems and locally developed solutions that may find application in new locales
Climate change will *not* be reversible	Try to slow migration to vulnerable coastal areas. Develop technologies for hardened infrastructure. Emphasize adaptation, mitigation, rather than reversal
Neither governments *nor* corporations will clean up legacy polluted sites	Find ways to contain toxins until future possible cleanups. Focus research on plants and bugs that metabolize the toxins. Design controls against unintended consequences of the plants and bugs

Source: Phillips (2014)

5.3.9 Risk Management

Risk management assesses and plans responses to collections of specific risks. With no overall envisioning of the future, risk management just aims to minimize risk, however defined, or to prudently take risks.

5.3.10 Speculative Fiction

Many adults now concerned with corporate and governmental futures spent their childhoods engrossed in the writings of Robert Heinlein, Arthur C. Clarke, and Isaac Asimov. Reading science fiction was, when the readers were youngsters, regarded as a fringe, geeky pastime. With today's staggering pace of technological change, it is hard to imagine a corporate leader surviving if she or he does *not* read science fiction!

Science fiction (and its more structured form known in France as *futuribles*) draws one plausible future, exploring its implications for purposes of food for thought, dramatic impact, or (often) a warning.

A good science fiction novel's strength is its fully drawn picture of the social and psychological impacts of a technological advance. Its weaknesses are that it reflects the view, however well researched, of a single author, and that a story is best told and marketed if it focuses on a single technological advance or event (a new invention, a nuclear war, etc.). The reality is that the future will bring many advances and events at the same time. Ours is a future where many new forces will interact, a future too complex and alien to be captured in a coherent story arc.

In sum, we have methods for looking forward that either...
Assume things will be the same,
   Just more so, or less so,
Or, allow for discontinuities,
   But still draw a single future, constrained by "conservatism of expertise,"
Or, pick a desired, plausible future
   And institutionally manage toward it,
Or, make an action plan that is robust to alternate futures,
Or, jockey for advantage in a more or less stable regime,
Or, allow a creative individual to envision a desirable (to him or her, anyway) future
   And let him/her take the risk of persuading the rest of us,
Or, discern the worst that can happen
   And try to avoid it
Or, viscerally engage us, to enhance our commitment to the future,
Or, create a variety of futures in order to increase our flexibility and resilience.

These are diverse ways of looking at the future! Members of a future-gazing team will have different implicit ways of conceiving of the future and the past. They will want different things from the future. They will have different expectations about the purpose and outcome of the project.

These differences must be made explicit and reconciled before any visioning exercise starts.

I should not omit to say that organizations also use much less rigorous methods of future gazing. They include hope, pessimism, gambling, conjecture, prayer, wishful thinking, and probably a host of others!

5.4   The Longer Run

It is interesting that in some focused areas, long-range planning is back. Questions of nuclear waste storage, environmental remediation, and archiving of digital information can be answered by plans that extend over multiple generations. These plans are very much worth making, but can still be obsoleted by newer scientific and engineering knowledge.[4]

Without doubt the next 35–50 years will present lots of "wild cards," which is what futurists call inconceivable events that materialize seemingly from thin air. Yet looking ahead 35–50 years is not atypical among foresight projects. The World Economic Forum published their look at the year 2050,[5] and even ventures a look at the year 2118.[6] The Millennium Project looks ahead to 2050 (Glenn and Florescu

---

[4]Or so I wrote in 2008. Phillips (2008).

[5]https://www.weforum.org/agenda/2017/04/this-is-what-the-experts-think-the-world-will-be-like-in-2050

[6]https://www.weforum.org/agenda/2018/01/heres-what-the-world-will-look-like-100-years-from-now/

2017). The XPRIZE Foundation is currently running a 20-year foresight project,[7] and Sokolov and Chulok describe a Russian effort (Sokolov and Chulok 2015) that takes on a 30-year horizon. None other than Winston Churchill bit off 120 years in his published musing on the world's condition in 2050.[8]

The main lesson from these efforts is that very broad generalizations about the decades ahead have a good chance of proving accurate. More specific predictions usually prove, well, awful.

The rate at which foreseen developments become real depends on funding. Funding forecasts (e.g., *R&D Magazine's* annual funding outlook) take a much shorter planning horizon, on the order of 2 or 3 years. This increases the uncertainty of science and technology (S&T) foresight, or alternatively forces the foresight project to use multiple scenarios, e.g., reduced funding/steady funding/increased funding.

5.5   And Even Longer: 500-Year Plans?

WIRED magazine's "Reality Check" column gathers experts' opinions about the direction and impact of technologies over a 200-year horizon. But these[9] do not represent a structured forecasting process and are meant mainly for entertainment and to open our eyes to new possibilities.

Five-hundred-year plans are an extreme example of the process described above and have been sighted in three settings.

- Timberlands West Coast Ltd. has begun work that conforms to a 500-year plan for sustainable management of New Zealand's West Coast Rimu forests. The plan is based on the 500-year growth cycle of the Rimu tree and was motivated by public opinion concerning the company's treatment of this unique resource.[10]
- The Booroobin Valley Project in Australia involves an Eco-Village with a 500-year plan. The plan is redolent of Utopian bravado.[11]
- A colleague of mine claims to have seen a 500-year operating plan drawn up by one of Japan's large conglomerates. It does not, he says, attempt to forecast what technologies will be creating value hundreds of years hence. It does describe ambitious expansion into new locations: the Moon, Mars, Earth's sea floor, the poles, and deep underground. It assumes that technological civilization will survive and that the Earth's enormous population will need and afford housing and amenities. It must further assume that corporate entities can maintain their

---

[7]http://www.iflscience.com/technology/what-do-you-think-the-world-will-look-like-/

[8]http://bigthink.com/the-voice-of-big-think/what-will-life-be-like-in-2050?

[9]Collected in Weiners and Pescovitz (1996).

[10]http://www.fao.org/Waicent/faoinfo/forestry/fhb/vol6-1/61-nw-e.htm

[11]http://www.maleny.net.au/sunweb/hintland/broobin.html

identities through the inevitable fall of governments and changes in social structures.

It is tempting to think that Japan's famous "patient [investment] capital" is patient because technology corporations there are conscientious about the very long future. Generally this is not true, though Japan has some of the world's oldest continuously operating business firms, and these firms do take a somewhat longer view of things than US firms do. An American electronics executive in Japan says, "Companies here know if they sell a product below cost for three years, U.S. competitors will drop out [leaving a clear field]."

(The Japanese speculative bubble of the 1980s showed little evidence of long-range planning, but let us lay that at the feet of the financial firms and the government.)

A Japanese colleague who made his career at a top-tier Japanese firm calls the idea of a 500-year plan "preposterous." Meaningful operational planning, he says, is only done on a five-year basis, and five-year plans are "no good" after 2 years.

Yet when I visited Japan frequently in the 1990s, there was considerable concern about a number of 30-year economic forecasts. The consensus of these was that in the year 2025 there will be simultaneous food and energy shortages, caused to a great extent by rising population and affluence in India and China, and by reaching the quantum limit of silicon semiconductor devices. In Japan, population will drop to 100 m from 120 m and will be much older. The older population will absorb public expenditure, leading to a weaker yen. Japanese technology forecasts do not see breakthroughs postponing this crisis much beyond 2025.

"Without breakthroughs in information, chemical, and biotechnologies, there is no future for Japan," said the chairman of the board of one top electronics concern. 30-year forecasts are influential and are taken seriously.

In any case, the notion of 500-year plans reminds us that in the very long run, values do change. The rise and spread of Christianity, Islam, and Buddhism are obvious instances. The Protestant Reformation in Europe illustrated that when changing values coincide with changing economic interests, big things happen.

5.6 Dispersion of Fortunes After 50–75 Years

History also shows that liquid capital aggregations disperse after a few generations. Fortunes in cash, stocks, and bank deposits are diluted by war, division among heirs, inflexible corporate cultures, and other reasons including generational effects—"First generation builds, second generation sustains, third generation squanders."

"Real" wealth, like the British crown jewels and the Vatican treasures, seems to endure longer, but the owners regard it as unthinkable to pawn these for investment purposes.

Some cute fiction has been written about secret, noncorporate money managers still running the deMedici fortune after all these centuries. But recent Swiss banking scandals (concerning Nazi deposits of stolen funds) show this to be fantasy. If the secret bankers really existed, they would be used for these kinds of transactions.

This implies there is a top limit to planning horizons using ROI as a criterion. It appears to be 50–75 years. After that, capital stops accumulating and starts decumulating. So beyond the 50- to 75-year point, what drives forecasting and planning? The drivers are ideas about the business environment, the natural environment, and sociopolitical change. To the businessperson, new values and new social organizations mean new markets.

Anticipating the existence of a market two hundred years hence is no guarantee of being able to connect with that market's customers in the context of contemporaneous social values and so on. Nor does it invalidate the 50- to 75-year rule and guarantee a steady increase in your stock price between now and then. To return to the point about farseeing Japanese firms: the existence of a 500-year plan is not necessarily an inspiration to investors.

5.7   Information Gain

Sometimes, rather than asking, "What do I know about the future," it's more helpful to ask, "How fast am I learning more about the future?" This isn't an easy concept, so let's come at it through examples.

Suppose you are a day trader. You want to take maximum advantage of early knowledge of the direction of stock prices. Seconds might matter (forget for the moment that algorithms can do this faster than you can), and you don't want to wait for your computer screen to refresh the ticker at a snail's pace. You pay for a faster Internet connection—or, as the institutional traders do, you move to a city that has bigger incoming Internet "pipes." You are then able to make more confident investments.

This means that gaining knowledge about the future becomes a matter of bits/second (or baud rate, or Shannon information gain).

Suppose you are a manufacturer and don't know whether you can get a reliable supply of the frammistats that are key components of the widgets you assemble. Tomorrow morning you sign a contract with a frammistat supplier that ensures your supply for the next 15 months. Now you know when and how many frammistats will arrive at your loading dock every day for a long time to come.

Suppose you are a general, and you have no notion of where the enemy will attack, and in what force. Your spies, scattered through enemy territory, manage to get messages through to you over a matter of days and weeks. They tell you, in bits and pieces, what weapons the opposing force has, how many infantry, what direction they will approach from, and the explosive power of their bombs. Each piece enables you to plan, and emplace, another aspect of your defense. You hope to have enough information, and hence enough preparation, on the day when it all hits the fan.

In short, as the examples show, if you are gaining information about the future at a rate faster than one day per day, you are ahead of the game!

Table 5.2 Attrition of new product ideas at each stage of the development process

Development stage	% of cancelled projects failing in this stage
Concept evaluation	19
Planning and specification	26
Development	37
Test and evaluation	14
Product release	5

Adapted from McGrath et al. (1992)

As everyone knows, few new product ideas make it all the way from lab bench to market launch. Companies cancel most new product development projects, for any number of reasons. In fast-moving consumer goods, about one idea out of 60 makes it all the way to the retail shelves. It's much worse in pharmaceuticals, where it takes 5000 experimental compounds to result in one marketable drug. It's better, however, in consumer electronics, where companies often just adjust the feature set of a failing new product idea, and manage to bring it to a viable market presence anyway.

One (real) consulting firm polled its clients as to how many new product development projects they spiked at each "stage gate" of the development process. Table 5.2 shows their results.

If 1 out of 60 product ideas bear fruit, we can say the probability of an idea becoming a marketed product is 1/60, or 1.7%. The consultant's table, however, lets us compute the "conditional probability" that a product idea will reach the market, *given that it has reached the planning/specification stage,* and so on for the remaining stages. This is the information gain quantification we are looking for!

But why were we looking for it? To launch a new product, a company needs to lock in supplier contracts, lease factory space, retain an advertising firm, and so on. All these imply not just current expenditures, but *cost commitments,* promises to continue paying for many future months or years. Our manufacturer's cost commitments reduce future uncertainty for the suppliers, the landlord, and the ad agency—but increase our manufacturer's anxiety. What if he's committed all these future payments and the product fails?

Let's ask the question a different way: how fast should the manufacturer commit costs? The answer is, *no faster than she gains information.*

My colleagues and I determined that some companies are good at this, and others are not. I'll skip the further mathematics that enabled us to draw Figs. 5.3 and 5.4 (and this means you may ignore the numbers in the figures), with data from real manufacturing firms.

To be sure, different kinds of product lines have different needs as regards capital equipment, factory space, and so on. Company B, whether through good management or through the luck of its product lines, does a much better job of synchronizing cost commitments with its rate of future-oriented information gain.

Figures 5.3 and 5.4 show a way of determining a company's "risk profile." If the company commits costs faster than it gains information, like Company A, it is a daredevil, embracing risk. Company B is risk-neutral, because its cost commitment

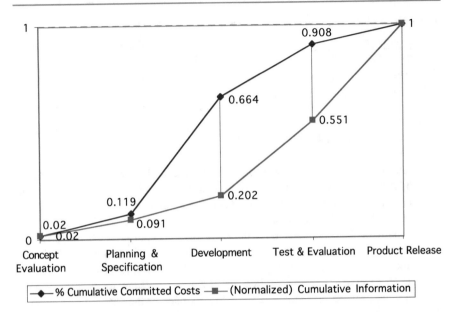

Fig. 5.3 Company A's rates of uncertainty reduction and cost commitment. Source: Author

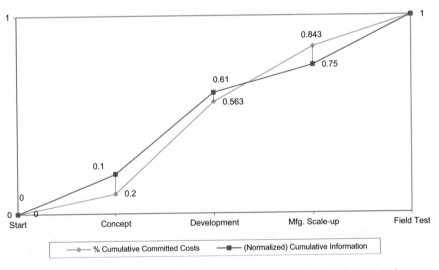

Fig. 5.4 Company B's rates of uncertainty reduction and cost commitment. Source: Author

and information gain curves match closely. A company that commits costs more slowly than it gains information is conservative, or risk-averse.

Obviously, the consultant's table, above, shows historical probabilities, not future probabilities. However, these probabilities had remained so constant over time that the consultants (and we researchers) were comfortable assuming the same probabilities would hold in the near future. Some situations may, in a similar way, lend themselves to the quantified comparison of cost commitment and information gain rates shown in the two pictures.

A Stanford University researcher (Baker et al. 2012) designed an index to measure the economic policy uncertainty that arises from taxes, government spending, and other policy matters. Brokers in Asia monitor the US economic policy uncertainty index, to decide whether to commit funds to long- or short-term investments (Thi Thuy Linh 2018).

In your childhood tree-climbing days, you gingerly put a little weight on a branch to see whether it would be safe to walk on it. Holding on to the tree trunk, you put progressively more weight on the branch, not to support your body, but *to gain information* about the branch's strength. Only after you were confident of its safety did you let go of the trunk and venture out onto the branch.

In this way, whether the situation lends itself to mathematical analysis or not, we can constructively ask ourselves, "How fast am I gaining information about the future?" Then we can refrain from committing our resources or our safety at a rate faster than our information gain.

5.8   All Communication Is About the Future!

Isn't everything we communicate really about the future? All the examples above—the investor, the manufacturer, the general—receive messages that guide their next actions. Idle chatter with friends and family serves to further bond them with you and with each other, bolstering your emotional state for tomorrow, and giving you confidence that you can rely on them in times to come. Even a report about archaeologists uncovering an ancient city helps us to understand where we've come from, and then to make more satisfying choices in the future. And that annoying colleague who always runs off at the mouth without saying anything of substance? He is teaching us to avoid him in the future, whenever possible.

An advertiser hopes her commercial message will induce TV watchers to buy her product today. There is really no communication that is totally about the past, or completely irrelevant to anything. All communication is, either in intention or benefit, about the future.

Key Takeaways

- Very long-range planning does not seem practical since fortunes tend to disperse within a 50- to 75-year time frame. Social values do change in the very long run. Yet some circumstances involving exploration, ecology, or stable business environment such as the touted "Asian great moderation" can justify planning ahead several decades or more.
- Speculations and scenarios based on knowledge of society and technology can aid our readiness.
- We should not commit our resources or our safety at faster than we gain information about the future.

References

Baker SR, Bloom N, Davis SJ (2012) Measuring economic policy uncertainty. Working Paper, Stanford University

Glenn J, Florescu E (2017) State of the future, v19.0. The Millennium Project, Washington, DC

Haynes KE, Phillips FY, Qiangsheng L, Pandit NS, Arieira CR (1996) Managing investments in emerging technologies: the case of IVHS/ITS. ITS J 3(1):21–47

McGrath M, Anthony M, Shapiro A (1992) Product development: success through product and cycle-time excellence. Butterworth-Heinemann, Stoneham, MA

Miles I, Saritas O, Sokolov A (2016) Foresight for science, technology and innovation. Springer

Phillips F (2008) Change in socio-technical systems: researching the multis, the biggers, and the more connecteds. Technol Forecast Soc Chang 75(5):721–734. https://doi.org/10.1016/j.techfore.2008.03.005

Phillips F (2014) Meta-measures for technology and environment. Foresight 16(5):410–431

Sculley J (1987) Odyssey. Harper & Row, New York

Sokolov A, Chulok A (2015) Priorities for future innovation: Russian S&T foresight 2030. Futures 80:17–32

Thi Thuy Linh T (2018) In: Yuan Ze University (ed) The asymmetric relationship between stock price and U.S. Economic Policy Uncertainty Index: Evidence on ASEAN-5 countries. Dissertation proposal

Weiners B, Pescovitz D (eds) (1996) Reality check. Hardwired

Seeing the Future: Ideas from Philosophy, Linguistics, and Sociology

6

> *Foresight activity promotes collective forms of learning that rely heavily on the cognitive attributes of foresight attitude.*
> —Jean-Philippe Bootz, Philippe Durance, and Régine Monti

6.1 Meaning, and the Future

What do we mean when we say something is meaningful? It can mean "suggesting something beyond the obvious," as in, "She cast him a meaningful glance." But we are pretending to be philosophers here, so we must ask what "meaning" means, for example, in Victor Frankl's famous book *Man's Search for Meaning*. To me, *meaning* means what an event, thing, thought, or emotion implies about a satisfying future for myself and for others.

To Frankl, finding *meaning* implied figuring out who you are, why you live, what you can learn from suffering, and above all, how to have *hope*. Hope for the future.

On the topic of artificial intelligence, Bernard Marr (2018) wrote, "But at the root of it all—the function which gives AI value—is the ability to make predictions." If Marr is right, the machines will have a lot of trouble comprehending *meaning* in the sense Frankl and I have defined it.

6.2 Epistemology

This branch of Western philosophy is concerned with how we know what we think we know, and why we think we know it. It puts forth three ways of knowing:

1. *Empiricism*: Knowing is possible only by interacting with the external world.
2. *Rationalism*: Knowing is possible via pure reason.
3. *Revelation*: Knowing is attained by direct contact with the divine.

© Springer Nature Switzerland AG 2019
F. Phillips, *What About the Future?*, Science, Technology and Innovation Studies,
https://doi.org/10.1007/978-3-030-26165-8_6

For my money, the boundary between the last two of these is but a thin line. Too many charismatic charlatans have believed they were in touch with the gods, but really just made stuff up. For me, it's empiricism all the way!

(Except when we're doing math homework. Proving theorems is rationalism, all the way.)

However well these three modes of knowing have served for apprehending the past and the present, they present a problem when looking at the future. There is nothing empirical to look at in the future, simply because the future has not happened yet. We fall back on a combination of rationalism and past-oriented empiricism, to extrapolate what the future might be like. And, of course, there are spiritualists who purport to rely on revelation to see the future.

We say "see" the future, but epistemology has more to say about the reliability of our senses, and the relation between what we "know" and the objects of our knowledge. Again, three schools of thought concerning seeing and apprehending:

1. *Epistemological Dualism* holds that the object "out there" and the idea "in the mind" are two entirely different things. One might have some similarity to the other—we're lucky when they do; in fact, our survival depends on it—but we shouldn't necessarily count on it.
2. Under *Epistemological Monism,* the "real objects" out there and the knowledge of those objects stand in close relationship with each other. Statements about physical objects are really statements about our sense data. Why? Because we are separated from the physical world and all we really have access to is our mental world.
3. *Epistemological Pluralism* is also known as *Relativism* or *Postmodernism.* It holds that knowledge is highly contextualized by historical, cultural, and other outside factors. Thus, a multiplicity of things affects the acquisition of knowledge: our mental and sensory events, physical objects, and various influences upon us that lie outside of our immediate control.

As regards the future, there are (as yet) no "real objects" "out there." So much for dualism and monism in future gazing.

Though many regard pluralism/relativism as unscientific—because it seems to lack objectivity—it has proven useful in planning. For example, will African elephants survive, or go extinct? It is largely up to humans. A ban on ivory exports would make ivory scarce and drive prices up on a black market, encouraging poachers. A limited legal market for ivory would cause scarcity with the same result.

A 2018 article in *Science* (Biggs et al. 2017) demonstrates that purely price-driven arguments have to be moderated by the values and cultures of the countries that are home to the elephants. Poachers have different alternative sources of income in the different countries, and elephants do more harm, less harm, or different kinds of harm to different kinds of crops in the different locations. Different human communities in Africa have different attitudes toward nature and conservation.

A purely economic analysis will not solve the problem, however "objective" it might seem. A facilitator who can elicit the local values and spur civil discussion among interest groups, the article notes, will have a far better chance getting local people to design their own way of ameliorating the problem and saving elephant populations.

6.3 Phenomenology

The future doesn't exist unless we assign *meaning* to it. For each of us, our sense of the future is bound up inextricably with the meaning in our lives.

Let's step back a moment to clarify what I did *not* just say. I am not saying the future doesn't exist if we don't think about it. That would be too simple.

Nor am I talking about *solipsism*, the idea that nothing in the universe exists until we, individually or collectively, conceive it. (There actually might be something to that argument, but it's above my pay grade as an amateur philosopher.)

It takes careful grammar to describe what "meaning" means! There is "small" meaning, which is the way we filter and interpret sensory experiences. This filtering and interpretation stands between a sensory input and the action we consequently take—an action, we hope, that enhances our survivability.

"Small" meaning is the domain of the branch of philosophy called phenomenology. You can see how phenomenology is future oriented. Sanz et al. (2011) write that "Understanding sensory flows and the derived emotional processes are strongly related to the anticipatory capabilities of the agents."

Then we have "big" meaning. It is the answer to "What is the meaning of (my) life?" Big meaning is your conviction that you have a purpose to fulfill. Again, obviously, future oriented. If you want to devote your life to helping the less fortunate, running a profitable business, or supporting your church, you have a purpose and you are looking to the future.

No purpose, no reason to think about the future. In the earlier chapter, I said the same about goals: No goals, no reason to think about the future. The two are a bit different; goals can be achieved, but purposes, or meanings, are ongoing. The future is bound up with both goals and meaning.

Who has no goals and no purpose? Even drug addicts have a goal, namely, to get their next fix. There is a pseudo-philosophy called *nihilism* that maintains life has no meaning. Certainly, the sheer number of absurd situations in everyday life might make one think so! Perhaps only sociopaths can be true nihilists, though even they strive for some kinds of self-gratification. Other self-styled nihilists dedicate their time to posting funny Facebook memes. That's a purpose, I suppose.

Physicist Carlo Rovelli (2018) is researching a quantum gravity theory in which time does not exist. This is so far from our everyday experience that we don't need to worry about it.

The Buddha said, "My mind is free from all past conditionings, and craves the future no longer." Yet he adopted a purpose, to convey the dharma to receptive ears. His spiritual heir, the current Dalai Lama, is a keen student of life's absurdities. He finds great humor in them, rather than dark despair, and continues to devote his energies to the happiness of others. No nihilism there.

On his hundredth birthday, asked about his future plans, comedian George Burns replied, "Plans? I don't even buy green bananas any more." Another aged comic claimed when he ordered a 3-minute egg, the waitress wanted payment in advance. Seriously and sadly, though, aged persons may feel useless. Without purpose, they are robbed of any sense of future.

Other than the disconnected elderly, and perhaps felons serving life prison sentences, only the truly mindless lack a sense of future (though they may survive until tomorrow).

Mindlessness is not peaceful. Lacking volition or purpose, the mindless are overly reactive—or maladaptively nonreactive—to outside stimuli.

To grasp the differences among mindlessness, mindfulness, and no-mind (*mushin*, in the Japanese Zen tradition in which I was trained) usually requires many years of practice. She or he who has achieved mushin still has a capability for intention, as Buddha and subsequent Bodhisattvas have demonstrated through lives of service.

Mindfulness is but one step in the Buddhists' eightfold path; it's limited without the other seven, but nice enough, and opens the door to the future. As management trainers worldwide climb on the "mindfulness" bandwagon, some adherents may be motivated to go beyond this "Buddhism Lite," and pursue more rigorous training.

Sanz et al. (2011) quote a sentence of Robert van Gulick. Though nearly incomprehensible, it is so beautiful that I must close this section with it:

> The phenomenal structure of experience is richly intentional and involves not only sensory ideas and qualities but complex representations of time, space, cause, body, self, world and the organized structure of the lived reality.

6.4   Linguistic Determinism and Technology Colonization

The San Diego Software Industry Council hosted an investor panel to discuss "What will Web 3.0 look like?"

"Why don't we know?" asked one venture capitalist. "Are we idiots?"

The VC was pandering to an in-group audience, with broad irony. The implied answer was, No, we're not idiots, we are successful, sophisticated investors, entrepreneurs, and scholars of the web!

But the truth is, yes, we are idiots. We are idiots because of *technology colonization*, and we fall for it every time.

In the mid-1990s the World Wide Web came along, and what did we do with it? We used it for push-publishing, for banner ads, and to sell stuff from web storefronts. In other words, we treated the WWW like an electronic magazine, or another television channel.

When we finally figured out the really new things the web could do for us—social networking, crowdsourcing, tweeting, distance learning, scientific collaboration, flash mobs, and blogging—we were so pleased with ourselves that we gave these a new, collective name: Web 2.0.

We might well have called it "YoucantdothatwithnewspapersorTV." It took a long time to figure it out. But in retrospect it looks, well, idiotic to have used the web to push text and photos in just one direction, with the only possible feedback being the "Buy Now" button. Why were we idiots? It was because the magazine and TV concepts, and the newspaper meme, *colonized* our perception of the WWW.

I first saw the phrase "technology colonization" in a 1995 *Wired* article authored by Barry Diller (1995). Media mogul Diller, then CEO of QVC (and later head of IAC/InterActiveCorp), urged us to resist "media imperialism," the tendency to define a new technology in terms of the old.[1] After all, he said, the automobile proved to be much more than a horseless carriage!

Those sneaky agents of technology imperialism, Texaco Theater, Milton Berle, and Ed Sullivan, held television back. Their shows were nothing but kinescoped stage plays and vaudeville. (Good plays and good vaudeville, but that's not the point.) Edward R. Murrow, George and Gracie, and Sid Caesar and Imogene Coca cut TV loose from the old model, showing us what TV could really do, as an eye to take the viewer into other people's homes, a vehicle for special effects, and a brave new world in which we watched ourselves watching each other.

The first automobiles were called horseless carriages because the carriage concept colonized the automobile concept, in the minds of producers and customers, limiting the auto's early usage modalities. By and by, users discovered that autos could perform functions that horse-drawn carriages could not, and society and infrastructure evolved accordingly. In fact, everything changed, from war (guns mounted on motor vehicles) to love (babies conceived in the back seats of '56 Chevys).

The colonization metaphor is apt. Think of the Jamestown colonists, who tried to build a little England in Virginia, an ill-conceived and ultimately fatal idea. In contrast, the Plymouth colonists—who actually called their place New England—mustered a bit of adaptability, shared a November meal with the locals, and thought "This isn't England, what can we make of it, hey let's build a park for the Red Sox, maybe recruit some eggheads and start a university or two." Theirs was America 2.0.

The Internet can do things that the post, the phone system, and town hall meetings cannot. But in the Net's early days, the post, the phones, and in-person meetings had colonized the Internet in the mind of the market at large. This may have been one reason for the dotcom bust. The only online businesses to survive the bust helped people perform familiar, cozy functions—auctions, romantic introductions, and job searches—more easily and quickly.

Of course, new net businesses sprang up in the post-bust recovery, but few people are using the net to do truly new things; to this day, everybody loves joining LinkedIn, but nobody really knows what to do with it.

Newsweek columnist Daniel Lyons agrees that journalists have used the Net "to do the same old thing. We take stories from newspapers and magazines and put them

---

[1]Sadeg M. Faris pushes a different definition of "technology colonization." He uses the term to mean a severe trade imbalance caused by the technological superiority of one of the trading partners. His is an important concept, but it is quite unrelated to Diller's.

on web sites. We publish books on Kindle. We put TV shows on Hulu." Writing in 2012, Lyons believed Apple's then forthcoming iPad, with its always-on Internet, would usher forth "phase two of media on the Internet," mashing together the styles of print and video journalism—but not, he says, until someone born and raised with the medium does the mashing. "Somewhere out there," Lyons wrote, "the Orson Wells of the digital age is [now] in grade school."

I hope to see research that sheds light on when Internet users (and users of any revolutionary technology) will throw out the colonizers. The colonizers are not malevolent people, but rather, old concepts of usage and functionality!

Some writers[2] think the 2001 Internet bubble burst because we ignored basic economic principles. On the contrary, it may have burst because we ignored colonization. If we can develop a rigorous understanding of colonization, it will be a better leading indicator of the Internet than the fevered gossip in Silicon Valley bars; we can answer the question (notwithstanding that some consider it already answered) of whether Meg Whitman, formerly of eBay, paid too much for Skype.

Taking the topic farther afield, writer Erik Davis (1998) echoes philosophers Martin Heidegger and Jacques Ellul in maintaining that our primeval myth stories and images are the primary drivers of our technology choices. That is, our myths colonize our technologies, too—not just vice versa.

Among Davis' most powerful points is that members of pre-technological cultures perceived themselves as integral parts of an animated world, in which each stone and tree harbored its own benevolent, mischievous, or malicious spirit. Nanotechnology and advanced electronics will once again complete the interpenetration of the magical and the technical, as today we move toward a techno-animated world where, as in a Disney cartoon, every teakettle will dance on mechatronic legs, sense with silicon/DNA circuits, and speak in a synthesized voice. For some years now, we've been able to buy Internet-connected smart refrigerators that monitor milk and order orange juice (Phillips 2005). In the coming year or two, the "Internet of Things" will take over our homes and workplaces much more completely.

Medical and biotech advances lead us to create Frankenstein monsters, simulacra of life that echo the mythical Golem.

It gets weirder: unfamiliar technology, in turn, colonizes the mythic imagination. It may do this in the form of demons, etc. Alien abduction hallucinations constitute an example—they are anxiety dreams about our own mutation. Davis quotes cybertheorist Michael Heim "We experience our full technological selves as alien visitors, as threatening beings who are mutants of ourselves and who are immersed and transformed by technology..."

Twilight Zone stuff aside, we may well look to our mythic past to divine the future. The stories we hear as children, which are already codified versions of mankind's ancient yearnings, inform our adult psyches. Consciously or not, then, we tend to invent things that realize those stories. Airplanes, of course. And many others, on the shelves of toy stores and the pages of the Sharper Image catalog.

---

[2]For example, Leibowitz (2002).

Diller believes we should be "convergence contrarians," willing to challenge conventional wisdom, able to explore other possibilities, treating a new medium on its own terms. Diller has been spectacularly right (Laverne and Shirley; Cheers) and spectacularly wrong. (In 1995, he said the arrival of cable did not signal the end of the networks. Not long after, a much-weakened NBC was sold to Comcast.) Is he right about colonization? I believe so.

Many suspect Web 3.0 will be the "semantic web." How long will we be hobbled by our understanding of Web 2.0, unable to see the true potential of the semantic web? What kinds of dippy, unimaginative apps will characterize the early semantic web—and more to the point, how soon will we let ourselves say goodbye to them, and move on to the good stuff?

Is "colonization by concepts" the same as mass delusion? No, it seems to more closely resemble a lack of imagination. Digital cameras, at first, seemed like a substitute for film cameras—the main benefit, saving on time and cost of developing negatives in the lab. But then people started to snap images they didn't want to keep for the long term—using them as short-term memory jogs—the modern equivalent of writing on the palm of one's hand with a ballpoint.

Then things went to a completely new level: My daughter got stuck while motorcycling in Thailand. She doesn't speak Thai, so she took digital photos of the broken chain on the bike, and the surrounding streets, then hitched a ride back to the cycle rental shop. She showed the proprietor the sequence of photos in the camera's viewer. He immediately understood what happened and where, lent her a new bike, and went to get the broken one. Digital photography as pictographic communication! Safe to say Kodak didn't anticipate such a thing.

Electronic computers, a faster alternative to mechanical calculators? That was the intention, but today far more people use computers to communicate—as I am doing with you at this moment—than to calculate.

I would hope that if we analyze enough such cases, we could extract a procedure for imagining the new and expanded uses of new technologies, and thus shorten the "colonization period." But before we get carried away with revolutionary fervor—kick out the colonizers!—we must consider the "Anchor and Twist" principle. This idea recognizes that it's easy to sell something new, but very hard to sell something revolutionary. So, for marketing purposes, a new technology must be presented (anchored) in terms of the old technology, but with a twist: "N is like N-1, but better, because it..."

Anchor and Twist, of course, perpetuates technology colonization. It is necessary in order to sell the idea to a conservative customer segment. But when selling to the "innovator and early adopter" segments (communication theorist Ev Rogers' terms for the more imaginative buyers), one can say, "Here's how we're going to anchor and twist with the slow adopters, but now let's talk about what you can really do with this gadget."

6.5   Technology and Society: Which Changes Faster?[3,4]

The eminent business historian Peter Drucker claimed that in the late twentieth century, technology was changing faster than society could keep up. In the years following his statement, people either quote him confidently or wonder whether it is still true. In 2011, I claimed the situation had reversed; society was by then changing faster than technology. Other researchers have since quoted me on that, and maybe taken it seriously.

In a way, the debate is meaningless. Look up "society," and you'll see sociologists define it as the way we live with each other and in our physical surroundings. That is, society includes the technological environment. It would be more meaningful to ask whether technology advances faster than *culture* changes—culture being the intangible aspect of society. We run into difficulty again, then, because there is no common yardstick to measure the speed of technical change (aside from some rather rough ones like the rate of patent applications) or the speed of cultural change.

Why did I write that society is now changing faster than technology? Government research grants are getting smaller and harder to win, putting a damper on engineering research and innovation. We are—probably—on the portion of the Kondratieff wave (about which more later) where industry is exploiting existing technology rather than investing in new tech.

Then there is the "technology fusion" phenomenon: Corporate R&D labs can grab (actually, license) an innovation from a distant country, and another one from a different company, and snap them together, like Lego blocks, with an innovation from their internal lab, and come up with something new and attractive to customers. Globalization and the Internet make this possible. There are so many innovations "out there" and available for combining that trying for breakthrough innovations within the firm simply isn't cost-effective.

Still other companies skip the product development task entirely, and just sweep up patents, buying them or licensing them from the originators. These companies' incomes flow from relicensing the patents at a markup, or from suing innovative companies that appear to be infringing on the patents.

All these trends have the effect of slowing technological innovation. The slowdown is obvious "to the naked eye" too, as we see Silicon Valley entrepreneurs developing apps that will call someone to pick up your dirty laundry. Hardly earthshaking innovation.

And culture? Everyone in my circle, and I'd wager in yours too, feels that our shared values are changing or disintegrating faster than ever before. Our beliefs are shifting—partly due to social media—and the patterns of who we communicate with are changing as well.

---

[3]This section adapted from Phillips (2011) and Phillips (2016).
[4]Drucker (1969).

6.6   And Yet Some Things Remain Cyclical

Aspects of the Smart City as a Service (SCaaS) movement have a "Back to the Future" flavor. Mobility-as-a-Service (MaaS) is a model for movement without ownership. Well, you say, that's what taxis are for, and public bike-sharing programs. For a monthly fee, MaaS will add faster and more flexible options, i.e., instant access to taxis, Ubers, buses, bikes, and so on.

One presumably very young blogger wrote of communication as a service. "Can you imagine," he enthused, "communication that's available to you everywhere, without carrying a phone in your pocket?" I sent a comment "Yep, phone booths." I suppose communication as a service will mean smart phone booths that know when you are near and route calls to whatever phone you are passing at that moment. Or perhaps he meant our phones will be implanted in our heads?

These fall into the "nothing new under the sun" category. There is another important cycle, however, that I call the Circle of Innovation.

6.7   The Circle of Innovation

New technologies create new markets and new ways of using products. The new markets and new usage modes spur social and organizational change, which results in new demands, and ultimately still more technological innovation. See Fig. 6.1.

The last link in this circular chain is usually recognized in terms of trends in corporate and government R&D budgets. But there are ways to look at this link that are more down-to-earth, more enlightening, and more fun. For example, continued crowding joined with increased affluence in Japan led to the bathroom appliance that

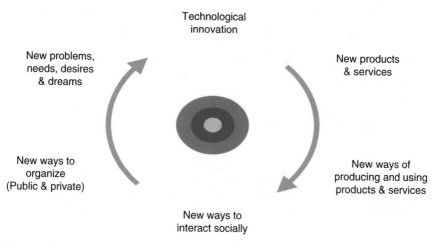

Fig. 6.1 The circle of innovation: Technological innovation self-reinforces via social change. Source: Author

combines the functions of commode, bidet, and warm air dryer all in one—ideal for small apartments.

The next section details another example, having to do with identifying scientists who happen to have the same name.

6.7.1   Individual Researcher I.D.'s (ORCID)

There are 185, yes 185, researchers registered in the Open Researcher and Contributor I.D. (ORCID) database who are named Wei Wang. Followed closely by 161 named Yang Liu. The rest of the top ten are also names of Chinese origin.[5] We can only guess how many more Wei Wangs engage in research but have not joined ORCID!

China's vast population and its recent increased wealth and openness (social changes) have led to a bigger footprint in international affairs (another social change), including world-class research.

Advances in information technology (IT) led to more international virtual research teams and wider access to scientific journals. This, plus the general globalization that is also enabled by new IT, raised research capacity in developing nations. Many more researchers from many more countries produce now work publishable in top international journals. Distinguishing among researchers having similar surnames (or names inconsistently transliterated into Western alphabets), never much of a problem heretofore, becomes an issue and an entrepreneurial opportunity. The universal researcher identifier is invented and promulgated. Publishing companies' author and reviewer databases now need to be modified to carry the extra data field "universal author identifier."

The earlier cozy research communities where (as in *Cheers*) everybody knows your name morph into a more impersonal but perhaps more productive enterprise.

In the ORCID example, summarized in Table 6.1, technical change led to new ways to use technology, which led to new organizational forms. These in turn created demand for new technological solutions. These, once provided, led to still newer usage modalities and a new round of social change in research communities. The wheel takes another turn.

The Broadway theater recognizes this interplay of social arrangements, technological change, and communication channels.

> A 2008 Tony winner for Best Revival, the swinging '60s farce *Boeing Boeing*... follows an American lothario living in Paris who's secretly engaged to three different flight attendants. But when the new, faster Boeing jet goes into service, the ladies' schedules get jumbled, and things turn turbulent as all three of them descend on his apartment at the same time, along with an old schoolmate who can't seem to keep his pal's cover stories straight.[6]

---

[5]Source: *Science*, 15 January 2016 • Vol 351 Issue 6270, p 213.

[6]www.nationalblackcalendar.com/event/boeing-boeing-swinging-60s-comedy-lands-in-coronado/

Table 6.1 Example: Individual researcher i.d.'s and the circle of innovation

Each event...	...turns the wheel.
Advances in ICT	Technological Innovation
E-journals; collaboration platforms	New products/services
More submissions from more countries to premier journals	New ways to use products and services
More international coauthorships. More authors with similar surnames	New social/professional interactions
Online conferences; global research teams; and bigger research communities	New ways to organize
Need to uniquely identify researchers with similar names	New needs and problems
ORCID and other identifier systems	Technological innovation; new product/service
Add fields to existing databases, to accommodate researcher i.d. number	New ways to organize

Adapted from Phillips (2014)

In still another real-world example, Apple's iPad and iPhone changed the way we work. Apple understands most iPhone users are bothered by the buzz of the smartphone and the constant checking of messages. The phones have become invasive. Technology distracts us from the things we should pay the most attention to—our family or friends, or something meaningful in our lives. To filter out useless messages and save the important ones, Apple introduced functions in the iWatch to make a different and better quality of life (Pierce 2015).

WIRED writer David Pierce asks, "Can technology fix a socio-psychological problem it [inadvertently] created with another piece of technology?" If so, it would be the perfect example of the Circle of Innovation. In fact, the iWatch uses your level of interest in the information, as demonstrated by your reaction to it, as a cue for the iWatch to prioritize, to get your face out of your tech. Apple has offered a feature called Short Look: An induced pulse on the wrist signals an incoming text message. The duration of the screen display depends on how long you cock your wrist and look at the watch.

The iPhone is an advertising medium. Apple has an incentive to maximize the minutes per day in which users are buried in the phone. Two things, though, moderate this incentive.

First, Apple is aware of the social damage smartphones are causing: "Cell phone zombies" wandering into traffic, families unable to talk to each other face-to-face, and so on. Enabling this kind of thing is not good for a company's image in the long run.

Second, by anticipating the negative (and positive) social impacts a company's own products will cause, the company gets an early jump on identifying the next big product opportunity. Whether we call it future gazing or technology assessment, the company is recognizing the Circle of Innovation, and profiting from it. Apple got first jump on connection-addiction countermeasure products, by virtue of assessing their own technology. By 2018, many other companies had jumped into that market, but did not enjoy first-mover advantage.

6.8   Psychology Is Real

People make up the "socio" in socio-technical systems, and if we wish to divine the future, we need to understand people better. Even the finance journals are now dealing seriously with "investor sentiment" as a driver of stock prices, and economists are moving away from neoclassical models toward "behavioral" models. Other examples:

- Population growth depends on fecundity, and fecundity depends on "selective memory" (our incomplete recollection of past pleasurable and painful experiences). This is simply because if the pain of childbirth were remembered accurately, few women would bear a second child.
- For lack of a few cents a day each, many thousands succumb to malaria, while we prefer to spend a small fortune rescuing one person in (for example) a cave-in.

I do not pass value judgments on these preferences. I just note that any rational futures our authors concoct will be stymied by human nature's bounded rationality and innate biases. Let's understand these, and forecast accordingly.

6.9   Can "Think Globally, Act Locally" Actually Work?

This is of particular interest because of an idea Hal Linstone often emphasized. We might call it the Linstone Principle. Hal's idea is that among the general population, the perceived importance and probability of an event drops off with its distance in time and space. The descending curve in Fig. 6.2 suggests that we give more weight to events that are in our own backyards, less about events on the other side of the world, more about things that happened today or will happen tomorrow, and less about what will happen 10 years in the future. Mother Teresa prefigured the Linstone Principle when she said, "If I look at the masses, I will never act. If I look at one, I will."

To Hal's conceptual abscissa (distance in time and space) we may add the scope of events, that is, whether the event affects a single person or masses of people. We'll support an individual orphan in Ecuador, or victims of a single tsunami in the Indian Ocean, but discount endemic poverty in Haiti. A single incident of abuse at Oprah Winfrey's South African girls' school[7] will move us, but we are cool to large-scale, ongoing domestic abuse in the USA.

Hal conjectured that heavy discounting of distant events was a survival trait through most of our history, but he noted that discounting hinders public support for action against objectively dangerous futures. He advocated "(1) using telecommunications to bring a distant event, crisis, or opportunity closer (as the

---

[7]https://www.telegraph.co.uk/news/uknews/1568410/Oprah-Winfreys-school-in-child-abuse-scandal.html

Fig. 6.2 The Linstone principle of discounting. Adapted from Linstone (1973)

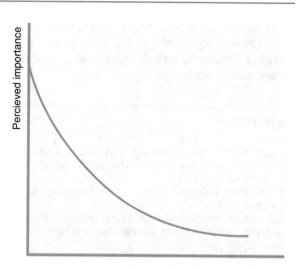

distant manned lunar landing was brought into our living rooms) and (2) using education to extend our perception outward" (Linstone 1973).

Even as we admit the ethical reasonableness of this recommendation (it was telecommunications that brought the Winfrey school case to our attention), we must note that others (and these are ordinary people, not just Don Vito Corleone) believe the pinnacle of morality is to reserve passion for people and events that are close, e.g., family above all others. I admit to you that I see both sides of this question, rationally and morally. (Some writers have urged, think about family first and save the world afterward. This argument, too, has pros and cons.)

Jenkins and Hsu (2017) have shown that if you give a decision maker a contextual story, for example, how he might feel next month if today he makes this choice or that one, he will discount the future to a lesser extent. Their finding is remarkable and counterintuitive, in that it shows better decisions are made for the long run when the question at hand is made a little *more complex*.

The next chapter discusses the ways our psyches deal with complexity. It will explain why some people take one side of this question and some take the other.

Key Takeaways

- While many forces are holding back the speed of technological innovation, culture—the intangible part of society, is undergoing faster-than-ever change. No matter which one changes faster, technology and culture have a reciprocal effect on each other and drive each other's development. This is the Circle of Innovation.
- By understanding the Circle of Innovation, companies can benefit from the social impacts caused by their own technologies.

- Technology is the more straightforward part of socio-technical systems. The other and more complicated part is people, with their bounded rationality and innate biases. Therefore, understanding human psychology and behaviors is key to predicting the future.

References

Biggs D, Holden MH, Braczkowski A, Cook CN, Milner-Gulland EJ, Phelps J, Possingham HP (2017) Breaking the deadlock on ivory. Science 358(6369):1378–1381. https://doi.org/10.1126/science.aan5215

Davis E (1998) Techgnosis: myth, magic and mysticism in the age of information. Three Rivers Press, New York

Diller B (1995, February) Don't repackage—redefine! Wired

Drucker P (1969) The age of discontinuity: guidelines to our changing society. Harper & Row, New York

Jenkins AC, Hsu M (2017) Dissociable contributions of imagination and willpower to the malleability of human patience. Psychol Sci 28(7)

Leibowitz SJ (2002) Re-thinking the network economy. AMACOM Division of the American Management Association

Linstone H (1973) On discounting the future. Technol Forecast Soc Chang 4:335–338

Marr B (2018, July 10) The economics of artificial intelligence—how cheaper predictions will change the world. Forbes. https://www.forbes.com/sites/bernardmarr/2018/07/10/the-economics-of-artificial-intelligence-how-cheaper-predictions-will-change-the-world/#29360f155a0d

Phillips F (2005) Technology and the management imagination. Pragmatics & Cognition 13 (3):533–565

Phillips F (2011) The state of technological and social change: impressions. Technol Forecast Soc Chang 78(6):1072–1078

Phillips F (2014) Triple helix and the circle of innovation. J Contemp East Asia 13(1):57–68

Phillips FY (2016) The circle of innovation. J Innov Manag 4(3):12–31

Pierce D (2015, April). iPhone Killer: the secret history of the Apple Watch. Wired. http://www.wired.com/2015/04/the-apple-watch/

Rovelli C (2018) The order of time. Riverhead Books, New York

Sanz R, Hernández C, Sánchez L (2011) Consciousness, meaning and the future phenomenology. AISB Workshop on Machine Consciousness, York, April 2011

Complexity and the Future

7

> In the future there's [sic] potentially two types of jobs: you tell
> a machine what to do, programming a computer, or a
> machine is going to tell you what to do.
> You're either the one that creates the automation or you're
> getting automated.
>
> —Y Combinator's Sam Altman

The protagonist of Karen J. Hasley's novel of Kansas (2014) moans that life on the prairie in 1919 is too complicated. Had Thea lived another 100 years, she would have seen this lede about an organization that commands far more resources than she could ever have brought to bear:

> [General Electric now] faces a different kind of challenge—a nightmare cash crunch that could take years to recover from. GE has been left in turmoil by years of questionable deal-making, needless complexity and murky accounting.[1]

What does it mean that a complicated world is getting ever more so? One good answer is that it is our job to keep our lives and our organizations from becoming so complex as to be unmanageable—as General Electric evidently failed to do. And to do that, we need to understand the nature of complexity, and understand how, psychologically, we tend to deal with complexity.

In this chapter, we'll attack those questions. We begin by painting our complicated world in terms of Multi's, Biggers, Smallers, and More Connecteds.

---

[1] http://money.cnn.com/2017/11/20/investing/general-electric-immelt-what-went-wrong/index. html

© Springer Nature Switzerland AG 2019
F. Phillips, *What About the Future?*, Science, Technology and Innovation Studies, https://doi.org/10.1007/978-3-030-26165-8_7

7.1   Multi's

In 1990, former DARPA Director Craig Fields observed that we were living in an age of eight "Multi's." The new enterprise environment was multi-product, multi-country, multi-culture, multi-company, multi-industry, multi-technology, multi-career, and multi-tasking. And so it remains.

But then, along came the World Wide Web. In 2019, we live with many more Multi's. Instead of working at one company, attending one church, and joining one lodge, we belong to many "groups" on Facebook and LinkedIn, attend multiple trade and professional society meetings, dabble with different forms of spirituality, and alternate between spinning, yoga, and cross-training.

Firms used to develop new products as a completely internal activity. The open innovation movement now means the firm's product developers accept inputs from many sources. Likewise, "commercialization," selling these new products for profit, is no longer the only way to bring their benefits to the masses. Products are now distributed by NGOs, directly by governments or their military forces, or by international agencies like UNICEF.

News came from the local paper and from NBC, CBS, and ABC. Now, we get news from blogs, email newsletters, Yahoo, Google, CNN, and Facebook.

We used to communicate with friends, family, and coworkers, either face-to-face or by telephone. Now, we routinely talk and text with teams scattered across the globe, and argue with strangers on social media. Business teams no longer show the easy agreement of a monoculture, but rather benefit from the multiple talents and approaches of team members from many cultures.

Multiple stakeholders with multiple objectives and agendas now insert themselves into every social or political problem—or, in the more civil discourses, are invited to participate. And as we saw with the African elephant preservation debate (in Chap. 6), multiple ideologies within and across cultures have to be considered.

Chapter 1 showed why multiple histories complement our usual notion of multiple futures.

The Multi's should be not only embraced, but encouraged and sometimes deliberately created, for example, multiple scenarios in planning, multiple decision bodies exercising checks and balances in politics, new alternative business models, and multiple methodologies in research studies.

Craig Fields urged his audience to recognize the Multi's and learn to tolerate them. I argue for more proactively embracing the Multi's in order to combine them in profitable ways.

The Multi's show we are living in a complicated world. It is likely to get more complicated. We should learn to anticipate the Multi's that are yet to come.

7.2   Biggers

I'll venture that we are living also in a world of Biggers:

- Bigger national economies that even strong political and corporate leaders cannot steer.
- Bigger perspectives on bigger problems (global warming, global terrorism, rogue asteroids), leading to new multinational and NGO actions.
- Bigger technological creations (dams, canals, etc.) having bigger impacts on people and societies.
- Bigger models for doing research and advising policy makers.
- Bigger companies growing as they merge and expand throughout a bigger global marketplace.
- A huge and growing volume of technical information.

7.3   More Connecteds

Each of these Multi's and Biggers displays increasing interconnections, both internally and externally. Though we could call these "multi-"interactions, I prefer a third descriptive category for today's world, the "More Connecteds." This helps us distinguish good from bad connections (an example of bad connections being "technology colonization," discussed previously) and paves the way for the discussion of systems modeling later in this chapter.

Our telecommunication and data networks, social networks, environmental problems, organizational structures, and regional economies become More Connected as each day passes.

7.4   Smallers

As Biggers become more omnipresent, so do Smallers. Microcircuitry now pushes the boundaries of quantum effects (and makes use of these effects, e.g., with the Josephson junction and for quantum computing), and nanofabrication represents, in effect, a convergence of mechanical engineering and analytical chemistry. 3-D printing brings us manufacturing on our desktops. Microsurgery and laparoscopic surgery allow invasive cures without traumatic cutting.

7.5   Multi, Bigger, Smaller, and More Connected, All at the Same Time

In a wonderfully illustrative table, Hal Linstone (1999) summarized how economic power is being pushed to the individual desktop and to microenterprises even as global conglomerates continue to merge and grow. (And the microenterprises, whether digital wedding photographers in India or textile designers in Ghana, are connected to suppliers and customers via the Internet.) History is driven by local languages, neighborhoods, ethnicities, and new states, Hal's table shows, even as Hollywood, McDonalds, and NAFTA consolidate a global culture.

This too is cyclical, as Hal and I have both written. Corporate forms go through cycles of conglomeration and decentralization, M&A and divestiture. "Democratizing technologies" like the sewing machine and the personal computer launch thousands of small seamstress and bookkeeping businesses—until someone with capital and marketing savvy rounds up the seamstresses and PC operators into sweatshops and call centers.

7.6   The Political Psychology of Complexity

A telling cartoon by Wiley Miller[2] depicts a crowd lined up for "Answers." Almost all follow the sign that says, "Simple But Wrong," being led directly to a cliff-edge, from which they fall to certain injury. Just a few follow the other sign, "Complex But Right," taking a winding path that leads who knows where?

The former president of University of Wisconsin at Madison, John D. Wiley, attacked[3] the staff of his state's biggest industry association for insisting that the answer to every question is "cut taxes," even as the association (and everyone else) watched Wisconsin's education system deteriorate. Wiley compared Wisconsin to other states, including neighboring Minnesota, and showed clearly that the most prosperous states do not have the lowest per capita tax burden.

This reinforces the more international analysis I did for *Review of Technology & Economic Development* in 2005: "The U.S., under its current small-government ideology, is seeing its sick go without costly drugs, and its K-12 education system decline. The Scandinavian countries—the most highly taxed and regulated on Earth, and the bane of small-government dogmatists—are highly innovative and entrepreneurial. That Denmark, Finland, Sweden and Norway are innovative, congenial places to live illustrates what this journal has long noted: Economic development is served by having a healthy, educated populace."[4]

---

[2]Gocomics.com/nonsequitur

[3]http://www.madisonmagazine.com/article.php?section_id=918&xstate=view_story&story_id=235966

[4]This paragraph also appeared in my *Social Culture and High-Tech Economic Development: The Technopolis Columns,* Palgrave, 2006.

I spent a day in 2008 with management guru Ian Mitroff, a colleague at *Technologi cal Forecasting & Social Change*. We discussed the Barack Obama–John McCain conversations at Rick Warren's church. Obama thoughtfully described complexities, while McCain told simplistic stories. McCain walked away with the day. Having evolved in simpler times, we are hardwired to love linear narrative—even when it's wrong. Interconnected complexities leave us behind.

Feelings, not facts, drive the electorate. That goes double when the facts are complex. If this was not obvious in 2008, it became shockingly so in the 2016 presidential election.

That day in 2008, I advanced the notion that insisting on simple answers in the face of overwhelming evidence of a problem's complexity can only be viewed as a mental illness. Another colleague voiced a counterpoint, that it's a common cop-out for academics to hide in complexities. Both are true—and all three of us agreed that it is the job of the thoughtful person to make a complex story understandable (as UW President Wiley did very well in his editorial) if we want large numbers of people to act.

If we don't, then only a handful of people will be willing to take action. The remainder will fall asleep, and it will be our fault.

Daniel Sarewitz wrote in *Slate*,[5] "Most scientists in this country are Democrats. That's a problem." Sarewitz cited a 2009 Pew Research Center finding that 6% of US scientists are Republicans and 55% are Democrats.

Sarewitz's conclusion that "a more politically diverse scientific community... could foster greater confidence among Republican politicians about the legitimacy of mainstream science" is one that scientists will ridicule. If cold, hard facts don't sway Republicans, a squishy meme like diversity—which Republicans already dislike—will hardly do the trick.

Sarewitz also missed an important big picture point: the same mentality that makes the conservatives of 2018 mean (in the word's original sense of "small") causes them to prefer simple answers—even obviously wrong simple answers like Iraqi culpability for 9/11—over the hugely complex and system-oriented answers that today's problems demand. They may have loved $e = mc^2$, but they'll never love complex climate dynamics.

Worried about our country's current political polarization, I had been reading and thinking about the psychology of twenty-first-century liberalism and conservatism. In a blog[6] building on Wiley's argument, I advanced the notion that insisting on simple answers in the face of overwhelming evidence of a problem's complexity can only be viewed as a mental illness. (This modern conservative malady is counter-point to the liberal academic's tendency to declare complexity and then hide within it, without ever reaching a decision.)

---

[5]http://www.slate.com/id/2277104. December, 2010.

[6]http://www.science20.com/machines_organizations_and_us_sociotechnical_systems/partisan ship_ simplicityseeking_and_maladaptation-85767 and http://consciousmanager.blogspot.tw/ 2008/09/followups-maladjusted-republicans-tax.html

My next blog went on:

> Last week I ventured that voters who cling to simple answers to hugely complex questions
> are mentally ill. It would be more proper to say mentally maladapted, according to former
> American Psychological Association executive director Bryant Welch. Welch's new book
> *State of Confusion: Political Manipulation and the Assault on the American Mind* notes that
> the world was once simpler, and we are still adapted to that world—not to this one. Moreover
> (and this is the main point of his book), we are ready, nay eager, to allow self-interested
> parties to manufacture false but simple explanations that we can buy into. Of course he really
> means only one Party, and its name starts with R and ends with epublican.

The book was reviewed in *Miller-McCune* magazine.[7] The reviewer punched
holes in Welch's thesis—and to be sure, a psychotherapist can't make a living curing
mental pathologies unless there are plenty of pathologies out there, and some
therapists are not above inventing new ones from thin air. But Welch's main
argument survives the review, holes and all.

Anyway, then came John E. Schwarz, Professor Emeritus of political science at
the University of Arizona, whose *Washington Post* op-ed[8] reasonably asked "Why is
tax-and-spend worse than borrow-and-spend?" (The former generally considered a
Democratic favorite, and the latter a Republican tendency.) Schwarz's reasonable
answer: It's not worse, it's better. It's not only more moral (because it doesn't take
food out of children's mouths) but it's also better economically.

Schwarz showed that job creation has been far greater under Democratic
administrations for the past 50 years. No, they were not all government jobs! They
happened because government invested in expensive and risky new technologies,
entailing R&D costs that no single company would pay for. The results were
commercialized, creating high-paying private sector jobs. Without increasing the
National Debt.

One commenter then posted:

> This political diatribe by Phillips does not contribute to a meaningful discussion of the
> implications of these one-sided political demographics on science policy. A good case can
> be made that ideology may be leading federally funded science in several politically sensitive
> areas. The issue is worth serious consideration, given the striking numbers claimed by Pew.

Diatribe! My posts were mild. . . . I posted again:

> My last post noted that a former executive director of the APA, a U. of Arizona Prof
> Emeritus, and an ex-U. of Wisconsin president have suggested that an aversion to complex-
> ity is maladaptive. When three people of such caliber speak, their proposition must be taken
> seriously. It needs to be investigated before any ill-advised rush to balance the ranks of
> science on ideological lines.

---

[7]http://www.miller-mccune.com/article/357. In fact, however, Hal Linstone had put forth the same
notion—that our minds are adapted to an earlier, simpler world—in Linstone (1996). (*Miller-
McCune* has now changed its name to *Pacific Standard*).

[8]http://www.washingtonpost.com/wp-dyn/content/article/2008/08/28/AR2008082802851.html

This is especially important as the scientific areas generating political controversy today are those involving extreme complexity—climate, genetics, stem cell research, human development, etc.

Because climate studies are only part of the broad scientific enterprise, though, I do not think Sarewitz makes a salient point when he says that climate science results "delivered by scientists who are overwhelmingly Democratic, are used. . . to advance a political agenda that happens to align precisely with the ideological preferences of Democrats." Republicans, for example, are strong advocates of national defense. And it was, after all, scientists—not shadetree mechanics or tea-party high school dropouts—who brought us atomic weapons, satellite battlefield imaging, and the stealth bomber. Likewise, Republicans emphasize law enforcement, and it was scientists who developed DNA evidence techniques.

Another commenter then challenged Sarewitz' statistics: "The Pew sample is drawn from AAAS members.[9] Is the AAAS membership representative of American scientists in general?"

Elizabeth Graffy of the US Geological Service posted a sensible comment:

> The right question is not what political party scientists. . . belong to; past research has already shown that there tend to be associations between different scientific disciplines and political worldviews.
>
> It is somewhat misleading—though undeniably provocative and interesting—to focus on the appearance of a correlation between scientists' political affiliations and the policy options on the table for any issue. . . . If there is any question about whether such a correlation really exists, then the right focus would be on the institutions for deliberation, not on the political parties of scientists.

As still another election approaches, and as it appears, so far, again to be characterized by extreme polarization and name-calling, I believe putting partisan psychology on the scientific table "is worth serious consideration." Even if I've done some of the name-calling myself!

Does answering every policy question, regardless of its nature, with "cut taxes" indicate an aversion to complexity? (There was a time when "answer every question with 'cut taxes'" might have been an unfair caricature. Today it is no caricature, but a well-deserved characterization.) Is it maladaptive? Mr. Welch of the APA argues on the basis of evolutionary psychology, and all evolutionary psychology is speculative. I should also mention that Prof. Schwartz is a Fellow of Demos, an organization advocating a "strong public sector." So, I'm happy to entertain arguments against my appeals to authority.

Simplicity-seeking is admirable, in moderation. All scientists prefer simple laws and elegant solutions—when possible. Sometimes simple laws and elegant solutions are not possible. Are there people who are immanently immoderate simplicity-seekers? And others who are not? Do the two types gravitate to different political

---

[9]American Association for the Advancement of Science.

philosophies? Are (1) yes, (2) no, (3) let's test it, and (4) it's wrong even to raise the question, the full possible range of answers? Which of the four do you subscribe to?

Let's allow cognitive scientist George Lakoff of UC-Berkeley—that hotbed of liberalism!—to have the last word. His is a compelling argument,[10] and it politely stops short of accusing anyone of maladaptation.

> We tend to understand the nation metaphorically in family terms: We have founding *fathers*. We send our *sons* and *daughters* to war. We have *homeland* security. The conservative and progressive worldviews dividing our country… are encapsulated in two very different common forms of family life: The Nurturant Parent family (progressive) and the Strict Father family (conservative).
>
> … We are first governed in our families, and so we grow up understanding governing institutions in terms of the governing systems of families.
>
> In the strict father family, father knows best. He knows right from wrong and has the ultimate authority to make sure his children and his spouse do what he says, which is taken to be what is right.
>
> Many conservative spouses accept this worldview, uphold the father's authority, and are strict in those realms of family life that they are in charge of. When his children disobey, it is his moral duty to punish them painfully enough so that, to avoid punishment, they will obey him (do what is right) and not just do what feels good. Through physical discipline they are supposed to become disciplined, internally strong, and able to prosper in the external world.
>
> What if they don't prosper? That means they are not disciplined, and therefore cannot be moral, and so deserve their poverty. This reasoning shows up in conservative politics in which the poor are seen as lazy and undeserving, and the rich as deserving their wealth.
>
> Responsibility is thus taken to be personal responsibility not social responsibility. What you become is only up to you; society has nothing to do with it. You are responsible for yourself, not for others—who are responsible for themselves. . . .

In contrast, says Lakoff, the nurturant family encourages children to read, think for themselves, be open to influences outside the family, and on those bases decide for themselves what is right and wrong.

In Lakoff's view, the strict father/nurturant family distinction determines how a person comprehends systemic causation—or doesn't. Systems involve many interacting elements and are subject to risk and uncertainty.

Conservatives argue that a cold winter invalidates scientific pronouncements of global warming. System thinkers, says Lakoff, understand that warming over the oceans evaporates more water, droplets of which blow over the pole (due to prevailing winds) and fall as snow, lots more snow than usual, over conservatives and liberals alike in the upper USA. "Empirical research," Lakoff continues, "has shown that conservatives tend to reason with direct causation and that progressives have a much easier time reasoning with systemic causation."

---

[10]https://georgelakoff.com/2016/07/23/understanding-trump-2/. As for the "not disciplined, not moral, and so deserve poverty" view of the strict father, see https://www.theatlantic.com/busi ness/archive/2013/11/your-brain-on-poverty-why-poor-people-seem-to-make-bad-decisions/ 281780/ for recent evidence refuting this view.

This is because strict fathers expect immediate obedience, and in its absence, punish the child "swiftly and directly," leaving the child in no doubt as to the reason for the swatting. For this reason, Lakoff concludes,

> Many of Trump's policy proposals are framed in terms of direct causation. Immigrants are flooding in from Mexico—build a wall to stop them. . . . The cure for gun violence is to have a gun ready to directly shoot the shooter. . . . [These answers make] sense to direct causation thinkers, but not those who see the immense difficulties and dire consequences of such actions due to the complexities of systemic causation. . . .

Yet a survivable future depends on our ability to apprehend, and act on, system complexity. To a system scientist, that dread word *diversity* means backup, survivable redundancy, and system flexibility. It has nothing to do with political correctness. Here's what a lack of diversity means:

> More than 75 percent of the world's food comes from just 12 plants and five animal species, making people across the globe vulnerable to what the report calls 'catastrophic breakdowns in the food system.' The United Nations' Food and Agriculture Organization has estimated that there is a one-in-20 chance per decade that heat, drought, and floods will cause a failure of maize production in both the United States and China, which would lead to widespread famine (Saldinger 2018).

This is serious. So, I will still maintain that people who cannot grasp the meaning of the passage above are maladapted to the modern world. The "swift" part of "swift and direct" implies an inability to deal with the longer-range future.

7.7 Complex Non-linear Systems

Complexity science, its early advocates hoped, would improve our ability to predict the future. So far, its principal contributions have been (1) to establish that prediction is even harder than we thought, and (2) to explain why it's harder. It's harder in biology, because each new insight in physiology uncovers even greater complexities. It's harder in economics, because where supply and demand used to reach an equilibrium (at a price agreeable to both buyer and seller), our economy is now in "permanent disequilibrium," largely due to today's spate of innovations and industry disruptions. The economy is fundamentally "out of control" (Kelly 1994).

The sub-field of complexity science called chaos theory is especially instructive. It gave us the well-known "butterfly effect," picturesquely asserting that a butterfly flapping its wings in Tahiti may cause a hurricane in Florida. It means that small causes can have big effects.

A great many physical, chemical, and engineering systems are described by differential equations (don't worry, I'm not going to show you any of those), and chaos theory tells us that a small tweak to one of the quantities in the equation can turn a smooth mathematical curve—and the smooth performance of the physical system it describes—into wild, unpredictable, and uncontrollable fluctuation.

These fluctuations are real, and have been seen in physical systems. Much of chaos theory's carryover into social systems has been unscientific reasoning by analogy, and poetic flights of fancy. Some of it is pretty convincing, though real instances of chaos in society and the economy have not been demonstrated. (A possible—emphasis on possible!—exception is in patterns of stock prices.)

7.7.1   Chaos in Society's Future?

Like "horseless carriage," chaos theory is poorly named. The word "chaos" is loaded with cultural baggage, popularly meaning the absence of God—the opposite of "order," which is the presence and beneficence of God. Recall the *Get Smart* television series, pitting agents of CONTROL (the good guys) against the minions of KAOS (the baddies). The scientific use of "chaos" implies nothing of that sort. The very onset of life in the universe may have depended on chaotic transitions. Yet, like the chaos of lore, we sometimes deny chaos, sometimes hide it, and sometimes deliberately create it.

The sales growth of a new product is supposed to follow one of the differential equations that is prone to chaotic transition. See Fig. 7.1. Knowing we couldn't definitively say whether chaos really occurs in new product introductions, Namwoon Kim and I asked (Phillips and Kim 1996), "Why, if chaos is there—and we suspect it is—have we not seen it?"

Most of the answers lay in the realm of organizational behavior. When sales begin to crash, companies withdraw the product from the market. The sales data series is then "truncated," as we data junkies say, at the date of product discontinuance. We don't have a chance to confirm chaos's presence. When daily sales appear chaotic, a sales manager will aggregate them into weekly or monthly figures, making for a smoother appearance, before making his presentation to corporate management. A chaotic sales trend would imply the sales manager's clever marketing tactics are not working. Who would want to admit that?

A cartoonist summed up this organizational psychology by picturing an intern presenting a full-wall projection of a graph like the one in Fig. 7.2. The manager demands, "What can we do about this?" The intern replies, "We could rent the office downstairs..."

But back to seriousness. The appearance of chaos might also be wrongly attributed to measurement error. Scientific chaos is "deterministic," meaning not subject to statistical fluctuation, but social scientists accustomed to using statistical tools tend to see variances from a regular curve as random measurement error. And measurement errors are almost always bigger in social science than in physical science.

Businesses may take the appearance of chaos as an opportunity. When energy prices became deregulated, and their variance over time accordingly increased, energy companies launched futures markets, making money by hedging the price fluctuations.

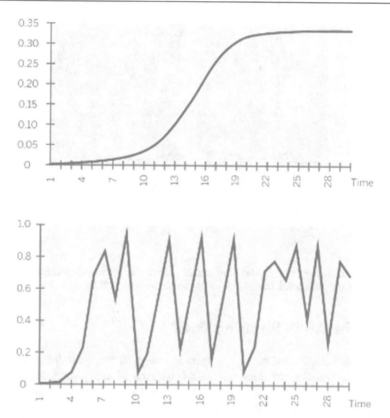

Fig. 7.1 In the top picture, the product gains ground in the market smoothly, topping out at 35% market penetration. At bottom, a small change in the curve's "parameter" causes chaotic fluctuation following initial smooth growth. What can a marketer make of this? (Source: Phillips and Kim[10])

Other businesses will accidentally create chaos, without really understanding it. The Lotka–Volterra equations, dating back to 1925, describe the fluctuating populations of competing species, or of predator–prey relationships. These equations say that the more rabbits there are, the more the well-fed foxes will breed, until too many foxes hunt too many rabbits and the rabbit population falls and the foxes go hungry and the fox population declines—whereupon the rabbits bounce back. Scientists have shown (Vano et al. 2006) that a four-species Lotka–Volterra system can fall into deterministic chaos. This means the rabbits may not always bounce back.

Now suppose four trading houses allow their artificial intelligence (AI) systems to issue buy and sell orders, based on the market's direction and on the actions of the other three AIs. In good Lotka–Volterra fashion, unrestrained by human judgment, these four AIs make oodles of money for a short time, then completely crash the markets. It has not happened yet, but it will.

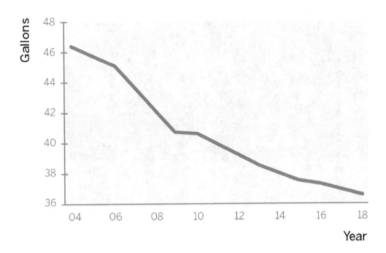

Fig. 7.2 Declining trend in soft drink sales, annual gallons per person. (Adapted from NASDAQ, http://www.nasdaq.com/article/coke-vs-pepsi-by-the-numbers-cm337909)

7.7.2   What About Managing, Then?

One is not anti-intellectual but simply realistic to say, "If you want to forecast in this environment—good luck!" One can but fall back on what my yoga teacher urges: keep a strong core and flexible limbs. Have a firm sense of what's essential (to you and to your organization) and what's dispensable.

No one enjoys a chaotic organization or a chaotic business environment. Generally speaking, no one can manage them, either. Hal Linstone and I have maintained that managers should direct their attention to *preventing* the runaway organizational or political complexity that could give rise to chaos.

In the classic management cycle, forecasts feed plans, which are implemented, controlled, evaluated, and modified. If current trends toward decentralization and empowerment in firms proceed toward autonomy of business units and employees, then what of management control? In the face of the mortgage meltdown, the Euro crisis, and disasters like Hurricane Katrina and the BP oil platform blowout—all of which researchers describe with the sanitary word "discontinuities"—a paradigm (the management cycle) will be dead, and it is not at all clear what role forecasting will play in any successor paradigm.

What we usually think of as the management cycle—plan, implement, evaluate—is a 100-year-old scheme (Fayol 1916) created for more stable business environments, but not for an age of discontinuities.

Yu-Shan Su and I drew new management cycles (Phillips and Su 2013), one each for the five regimes under which firms will operate: stable, chaos, edge of chaos, entrepreneurial, and disaster. We forecast that managers in the future will move their companies seamlessly from one of these cycles to another, as the situation demands.

Some companies are, or have been, almost there. Continental Airline and Toyota both had entrepreneurial beginnings. Continental enjoyed stable growth from 1934 until the US airline deregulation in 1978. Chaos characterized a failed merger attempt (with Delta) and boardroom shenanigans through 2001, when disasters took over: the 9/11 attacks, Hurricanes Rita and Ike, and a runway crash in Denver. Continental rebounded and merged with United in 2010.

Toyota's famed "Toyota Way" encompassed the five management cycles and emphasized organizational learning. This constructive way of guiding a complex organization was abandoned in 1999 in order to pursue faster growth. The result was quality and recall crises, handled clumsily, and loss of market share. Toyota had emphasized HR procedures and quality of people, but ultimately failed to trust their people, continuing an overcentralized control that lost touch with operations and the market. (In a new book (Greulich 2017), a 30-year veteran of IBM explains how IBM went through the same kind of unfortunate transition.)

Toyota's CEO had once said: unless we enhance quality today, we cannot hope for growth in the future. Satirists at *The Onion* joked that Toyota would recall all 1993 Camrys, because that model was too reliable and the company judged it was simply time (around 2015) for Camry owners to buy a new car. General Electric shed its venerable light bulb business after introducing LED bulbs, because the new bulbs last 10 years, people are consequently buying fewer bulbs, and the bulb division was no longer profitable. One imagines executives face-palming, realizing that when their product becomes long-lived and reliable, it's great, but it's high time to think of newer and different product lines.

Adaptive systems are flexible. Does "flexible company" mean a "no planning" company? Of course not. But the best balance of flexibility and efficiency is just now becoming quantified (Phillips et al. 2019),[11] and in many companies, global hypercompetition and remnants of the shareholder capitalism ideology dictate that cost-cutting still takes precedence over flexibility.

7.8 One Man's Record of Predictions: Right, but Not Right Enough

System theory has progressed beyond the old definition of a system as a fixed set of entities (nodes) and interconnections (arcs), as Fig. 7.3 suggests. We now know that *the system is the set of generative rules* that govern the birth and death of nodes and the evolution of their interaction.

---

[11]See also Isobe et al. (2004).

Fig. 7.3 A system may look like this at one point in time. But what governs birth and death of the nodes and arcs? (Source: Author)

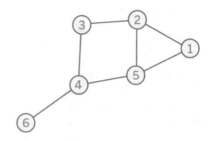

In 1996, with this new understanding of systems, I ventured a few predictions. Let's see how they turned out:

1. *The current system of corporate stock ownership will disappear within 30 years.* The growing importance of entrepreneurial technology companies led to this conclusion via the following reasoning.

In an era of relationship marketing, the stock markets are among the last still to operate on the old arm's-length-transaction model. Entrepreneurs are famously at odds with venture capitalists, because entrepreneurs risk everything for their enterprise, and give it their personal "all." Venture funds have a legitimate responsibility to their investors, that is, to steer the fund to the highest return. The market is too liquid. Entrepreneurs need a commitment from investors that matches their own, and they are not getting it. I noted that Clinton and Gore's proposed reform of capital gains taxes to reward less volatile investing was a small step in the right direction, but was diluted after Republicans won control of Congress in 1994.

The 30 years are nearly up, and my prediction seems off the mark. Yet the IPO market was down in 2018, and public companies like Dell are going private. It might have been more accurate to predict, in 1996, that the era of the publicly traded company will come to an end.

A company's "generative rules" are its basic culture, procedures, and knowledge. These may persist as the firm experiences mergers, alliances, and spin-offs. They inhere in the firm's entrepreneurs, employees, and technologies. They propagate as groups of employees move to other companies.

No mechanism exists for investors to bet on entrepreneurs or technologies; the current stock system allows them to bet only on companies. Thomas Edison failed in a dozen ventures before commercializing the light bulb and phonograph. Had it been legal, an investor would have been wiser to invest in Edison rather than any of the particular ventures. As growth technology firms attract a greater fraction of the world's stock capitalization, I said in 1996, this fact will become more obvious and weighty. The second 1996 prediction was:

2. *Much dislocation will follow as new investment instruments are invented, approved, and marketed, and current stockholders cash out and reinvest.*

Some universities are waiving tuition in favor of taking a fraction of graduates' salaries for a certain number of years. This admirably shifts risk from students (who were previously burdened by student loan repayments, whether employed or not), but smacks uncomfortably of indentured servitude and a violation of the 13th Amendment to the US Constitution. The 2000s was the decade of "financial innovation," giving rise to all kinds of horrors, including hedge funds and collateralized debt obligations. Global capital mobility under the "free trade agreements" has reduced stock holding periods. Let's award prediction #2 a "pretty good."

3. As an intermediate step, because individual technology investors will not tolerate the insiders' market in initial public offerings for much longer, *IPOs will be securitized and marketed to small investors, possibly in the form of mutual funds.*

We now have crowdfunding, via Indiegogo, GoFundMe, and a host of others. These are for pre-IPO offerings, so prediction #3 was half right. Interaction with another trend—the Fed keeping interest rates near zero—led to an unfortunate consequence of investment liberalization. An "ever-expanding bubble in startup valuations, fueled by an ever-expanding pool of increasingly less qualified investors. . . . [Due to] pitiful or negative interest rates. . . people with a lot of capital will pay almost any price for the chance to earn a meaningful return" (Popper and Lopatto 2015).
More recently, I predicted. . .

4. . . . and, as I confessed in Chap. 4, promptly forgot that I'd predicted 5 years in advance, that *a Donald Trump, or someone equally radical on the left, would take the US presidency.* I am as prone as anyone to indulge in denial, when a prediction is extremely distasteful.
5. With anti-immigrant sentiment becoming more visible in America, and considering that so many American innovations (80% of all recent US patents) were granted to immigrants—and with a March, 2017 *Forbes* headline asking, "Is Brand America Tanking?"—I ventured that *an America that is less attractive to entrepreneurial migrants would be a less vibrant innovator.*

Sure enough, early in 2018 Bloomberg News Service dropped the USA[12] from its top ten innovative countries ranking. Short-term forecasts sure are easier than the farther-out ones.

I have been generous in grading my own predictions, but the point is that even to the extent they were correct, they were not correct enough to guide actions that would have benefited me, or anyone who might have been foolish enough to listen to me. In the Greek myth, Cassandra delivered true prophecies that no one believed.

---

[12]https://www.forbes.com/sites/andrewlevine2/2017/03/28/is-brand-america-tanking/ *and* https://www.bloomberg.com/news/articles/2018-01-22/south-korea-tops-global-innovation-ranking-again-as-u-s-falls

The Greeks and Trojans were wise not to believe. Cassandra's visions may have been true, but like mine, not true enough (mostly) to do anyone any good!

7.9   A Final Word on Systems

The late great storyteller Poul Anderson[13] described the future in terms of interactions of old and new categories—not as the straight-line progress of separate categories. Stanford economist Paul David (1991) described the effects of the computer and the dynamo on the twentieth century. These "superbasic" (my term) technologies give rise to multiple industries, have a central role in economies over a period of 75–100 years, and are slow (60–70 years) to achieve their full effect on organizational productivity. *Business Week* declared the 21st to be the "biotech century." But nanomachines, superconductivity, and even anti-gravity are just as likely to be the superbasic technologies of the next century, and *they will interact.* Writers and strategists are already thinking about how artificial intelligence and blockchain will combine to create the business future.

The number of possible combinations of *n* technologies grows faster than exponentially as *n* increases. The future will be complex indeed.

Modern system theory teaches that systems in chaos die quickly. Systems near equilibrium die quickly or slowly—because their environments change. Systems "on the edge of chaos" can survive and evolve—but not necessarily in predictable ways. Our students' students may find it quaint or even incomprehensible that we of the twenty-first century were so preoccupied with predictability.

Key Takeaways

- We are living in a complicated world of Multi's, Biggers, Smallers, and More Connected, while our psyches are prone to simplicity regardless of its potential wrongness. However, our ability to understand and deal with complexity is essential for a survivable future.
- Aversion to complexity seems tied to America's political polarization. One personality may hide behind perceived complexity of a situation, and never arrive at a decision. Another personality may insist against all evidence that a complex situation is simple. Both responses are maladaptive.
- As the business environment unceasingly evolves, forecasting is becoming increasingly difficult, as it can be hard to generate right-enough-to-be-useful predictions. Therefore, instead of being exclusively concerned with forecasting, managers should work to prevent the potential chaos arising from organizational and political complexity.

---

[13]Poul Anderson (1996). See also Mitsubishi Research Institute (1991).

References

Anderson P (1996) All one universe. St. Martin's Press, New York

David PA (1991) Computer and dynamo. Stanford University Center for Economic Policy Research, Reprint No.5

Fayol H (1916) Administration Industrielle et Génerale. Bulletin de la Société de l'Industrie Minerale Fifth Series, 10(3):5–162

Fields C, R&D Consortia (1990) Speech at IC^2 Institute of the University of Texas at Austin

Greulich PE (2017) Think again. MBI Concepts Corporation

Hasley KJ (2014) Smiling at heaven. Createspace Independent Pub

Isobe T, Makino S, Montgomery DB (2004) Exploitation, exploration, and firm performance: the case of small manufacturing firms in Japan. (Institutional Knowledge at Singapore Management University Research No. 10–2004)

Kelly K (1994) Out of control: the new biology of machines, social systems, and the economic world. Addison-Wesley, Reading, MA

Linstone HA (1996) Technological slowdown or societal speedup—the price of system complexity? Technol Forecast Soc Chang 51(2):195–205

Linstone HA (1999) Complexity science: implications for forecasting. Technol Forecast Soc Chang 62:79–90

Mitsubishi Research Institute (1991, July) Towards the next century, Tokyo

Phillips F, Kim N (1996) Implications of chaos research for new product forecasting. Technol Forecast Soc Chang 53(3):239–261

Phillips F, Su Y-S (2013) Chaos, strategy, and action: how not to fiddle while Rome burns. Int J Innov Technol Manage 10:6

Phillips F, Chang J, Su Y-S (2019) When do efficiency and flexibility determine firm performance? A simulation study. J Innov Knowl 4:137–147

Popper B, Lopatto E (2015, Dec 29) Silicon Valley is confusing pseudo-science with innovation. The Verge, https://www.theverge.com/2015/12/29/10642070/2015-theranos-venture-capital-tech-bubble-disruption

Saldinger A (2018, January 18) Climate, cybersecurity top list of global threats in new report. https://www.devex.com/news/climate-cybersecurity-top-list-of-global-threats-in-new-report-91899

Vano JA, Wildenberg JC, Anderson MB, Noel JK, Sprott JC (2006) Chaos in low-dimensional Lotka-Volterra models of competition. Nonlinearity 19:2391–2404

Analytics and the Future

> *Science has traditionally been seen as a driving force for*
> *technology, but the inverse process is equally important.*
> —J.P. McKelvey (1985)

8.1   Growth, and Tipping Points

You start a company, and add one new employee every day. Assuming you hire someone in the first day, the number of employees in day d is $1 + d$, the "1" being you. Your company's headcount equation is $E = 1 + d$. This is the equation of a line, so a graph of your employment growth looks like Fig. 8.1.

Perhaps you've heard this riddle: a tiny algae colony in a pond doubles in size each day, and will completely cover the pond in 30 days. On what day does it cover exactly half the pond?

Rather than adding a fixed amount w to the population every day, the algae are each day adding a multiple m of the existing population. In the riddle, $m = 1$. Each day the existing population E gets bigger, so each day mE gets bigger. If the starting population is p, we have

$$E = p(1 + m)^d, \text{ or since } m = 1, E = 2^d p$$

d is an exponent, so the growth is called exponential growth.

One point of this instructive riddle is that exponential growth is really, really fast (see Fig. 8.2). The riddle makes a second point, equally important but often overlooked: the pond has boundaries. There is a limit to the algae colony's growth.

And a third point is only implied: what happens on day 31? One likely outcome is that the algae cover will prevent oxygenation of the water, killing fish and disrupting the pond's ecological balance.

© Springer Nature Switzerland AG 2019
F. Phillips, *What About the Future?*, Science, Technology and Innovation Studies,
https://doi.org/10.1007/978-3-030-26165-8_8

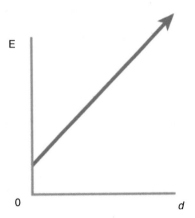

Fig. 8.1 Linear growth. Source: Author

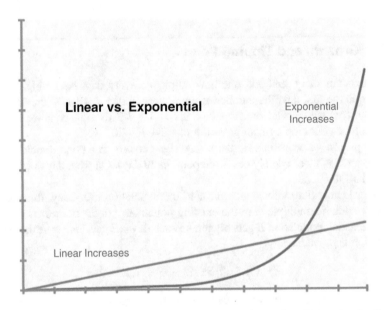

Fig. 8.2 In the long run, exponential growth is much faster than linear growth. Source: Figure adapted from www.conanstevens.com

Furthermore—if you will allow me still a fourth point—the time between zero and the next moment that Fig. 8.2's two curves cross has been called the "region of disappointment." A manager not familiar with the math may accuse the engineer, "We could built more robots by now if we just assembled two of them every day ourselves. But no, you wanted to make robots who would build more robots, and it's slower, don't you see!" The engineer is aware of the truth—that if each robot builds two more robots every day, and *those robots* build two more every day... production is slower than linear in the short run. But much, much faster in the long run.

The answer to the riddle, by the way, is that the colony covers exactly half of the pond on the 29th day.

8.2 Trend Extrapolation: S-Curves

The pond has a definite carrying capacity, just as the market for a product is not infinite, and just as there's a maximum speed for an aircraft (before it becomes a spacecraft). In Fig. 8.3, K denotes the carrying capacity or market potential. Though near the zero point the two trajectories shown in the figure both look like exponential growth, the lower curve "recognizes" the effect of the limit K, and levels off as it approaches K. The upper, exponential curve ignores the carrying capacity and keeps growing beyond the edge of the pond—an impossibility, unless the algae grow legs!

The lower curve in Fig. 8.3 is called a logistic curve, demonstrating "logistic growth." The logistic curve is one of a family of "S-shaped" curves that have successfully modeled the diffusion of innovations, the market penetration growth of new products, the life cycle of industries, and, of course, the growth of plant and animal populations in circumscribed ecosystems.

Elaborations on S-curves (all still S-shaped) variously take into account slower growth when there is organizational resistance to change; faster growth when prices

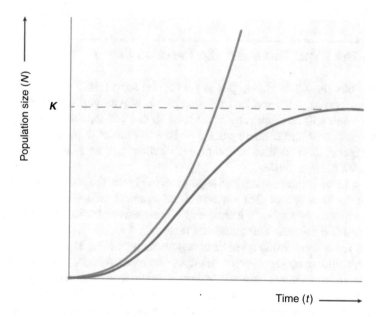

Fig. 8.3 Exponential vs. logistic growth. Adapted from https://swh-826d.kxcdn.com/wp-content/uploads/2011/12/exponential-vs-logistic-growt.jpg

are reduced or when buyers try to "keep up with the Joneses," imitating their neighbors buying habits; and even faster growth under the "network effect," when buyers influence and are influenced not only by immediate neighbors but by distant contacts on social media. Still further variations account for repeat buying and multiple generations of products like iPhone 3, 4, 5, and 6.

Thoughtful professionals prefer S-curve growth models—not because we like to throw cold water on the party, but because limits to growth are a reality. Everything that grows eats resources, and when the resources dry up, growth must slow or stop. Mobile and intelligent creatures facing population pressure can move to a new region and start again—as Europeans colonized the New World and as humans now look into living in space and on and under the sea—but this does not change the fact that growth has slowed in the origin region.

People who loudly declaim "exponential growth" have something to sell, or something to hide. In the early stages of, for example, artificial intelligence, its proponents are not lying when they say "exponential," because at that time exponential and logistic growth look similar. What they are hiding is that exponential will turn into logistic, pretty soon.

Your author worked at General Motors Research Laboratories' math department when the famous *Limits to Growth* book first came out. I was assigned to replicate the book's model. It took me a while to realize that GM's board and top management could hardly care less about this interesting model, but were hoping only that I would disprove it. What was then the world's largest company could not afford to believe there are limits to growth.

But there are.

8.3   The Hype Curve and the Reaction Curve

"I had to buy the whole album, just to get the one song I like." You've heard this complaint; maybe you've voiced this complaint. Maybe you downloaded (don't worry, I won't tell) an illicit copy of the song so that you wouldn't feel ripped off. Record companies, rather than update their business model to make customers and artists happier, instead tried to bring legal action against kids who posted or downloaded bootleg music.

Then start-up companies devised legal ways to give us the music we want, when we want it. Start-ups of that ilk are called *upstarts*, and the defensive traditional companies *incumbents*. Incumbents almost always try to backpedal against innovations that threaten their traditional ways.

In the same way, inside a business organization or even in a family, insecure individuals oppose change. Parents want a more modern house; kids whine about having to move to a new school. Managers bring in new order entry software; employees flood the email server with messages to the effect that the old software was better, the new program is too hard to learn, the manager is stupid, I'm not going to use the new one and you shouldn't either, and so on.

Resistance to change is a powerful force.

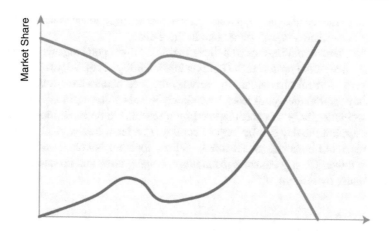

Fig. 8.4   Reaction curve: Effect of the hype curve on the incumbent. Source: Author

Yet sometimes the new technology really is faulty—rushed to market before it was fully tested. Or, like a fast-food meal that tastes okay but looks nothing like the picture on the menu, a new tech may perform adequately but still fall short of the vendor's advertising hyperbole.

Figure 8.4's lower curve, emerging from the zero point, is called a hype curve. It pictures the situation of the previous paragraph. The small peak is often labeled "inflated expectations." It's followed by a "trough of disillusionment," when disappointed customers stop buying. An "enlightenment" stage reflects successful bug fixes, and the final, rightmost stage finally suggests widespread, productive use of the new tech.

In mirror image to the upstart's trajectory (bottom curve), the incumbent (top curve) experiences stages of myopia, denial, alert, reassurance, and unwarranted complacency, followed by panic and resignation.

An example from my own industry experience in one of the USA's two oldest market research companies: When upstart companies started collecting market data from bar code scanners, we (the incumbent) didn't notice at first (*myopia*). Then we thought it was all flash with no substance. (This is *denial.*) The new technology was glitzy and sexy; the client companies were all over it like kids in a candy shop, and we started losing customers to the upstart firms. (We were now *alert* to the threat.) The upstarts showed that they could not guarantee high-quality data, nor process it intelligently, and customers began to return to us. (*Reassurance and unwarranted complacency.*) The upstarts fixed bugs and improved programmer training, and customers decided anyway that they liked cheap and fast data better than correct but slow data. We *panicked* and finally *resigned* ourselves to a new era in the market research industry.

Because the new and the old technologies share the total customer base, the two curves cross when the upstarts and the incumbents each hold 50% of the market.

This point is reached after all the psychological, marketing, and technical gymnastics just described—that is to say, rather late in the game.

There are theoretical reasons to believe that this 50–50 point signals an irreversible dominance of the upstart tech (Phillips 2007; Phillips et al. 2016). But because this decisive moment arrives late in the war, the incumbents have lots of time to believe they might win—or at least, lose slowly—and so they fight all the harder.

New technologies, when accompanied by a reasonable business model, usually take the day, but the day can be long in coming. (To learn how a great technology with a miserable business model totally failed, look up Nikola Tesla's wireless electricity tower. Clearly amazing technology; equally clear that no one could ever make a penny by offering it.)

8.4   Price–Performance Curves

Intel Chairman Gordon Moore's prediction that density of transistors on a chip would double every 18–24 months—at the same cost!—has held true for decades, driving price reductions in electronics products, and making our computers and phones affordable. It's been estimated that if the price of automobiles had declined at the same rate, cars would cost a nickel today.

As consumers, we have been extraordinarily lucky, at least as regards electronics. Managers of chip-making companies likewise have been extraordinarily lucky, to be able to run their businesses "by the numbers," basing their pricing strategies on predictable cost reductions.

Have we (and they) reached the limits of Moore's Law? Circuits are now so tiny that heat dispersal and quantum interference are nearly insurmountable problems. Though there are some work-arounds, it looks like the end of the road. As I mentioned in Chap. 1, some companies' plans presume Moore's Law will hit a brick wall around 2025.

This will not crash our economy, for reasons I'm about to explain. You might want to shift your investments, however, from electronics to genomics. Or solar power.

Previously costing millions, it now costs about one thousand dollars to sequence a genome. This price decline is even steeper than Moore's Law, and there's no end in sight. New businesses are popping up; there were one hundred VC deals in genomics in 2015 with investors pumping one billion dollars into start-ups each quarter.

Another market showing supra-Moore price declines is solar energy. The cost per kilowatt-hour of solar electricity is nosediving.[1]

The chip, gene, and solar panel illustrations are examples of *price–performance curves*. What are the implications of these dramatic, and possibly unprecedented in history, cost trends? For one thing, it used to be the marketing department's job to find excuses to raise prices. (New! Improved! Gets your clothes cleaner!) Now, it's their job to schedule price *decreases*.

---

[1]http://www.solarcellcentral.com/cost_page.html; https://seia.org/solar-industry-data

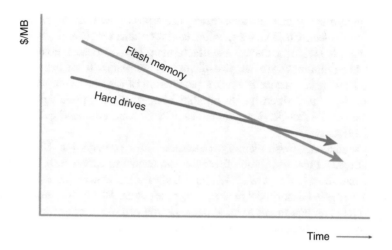

Fig. 8.5 Costs of two memory technologies decline, but one declines faster. Source: Author

The trends also mean that companies' plans need to be based on *technology substitution*. Figure 8.5 shows the costs of flash memory and hard disk memory both declining. But hard drives, at one time cheaper than flash memory on a cost per megabyte basis, were getting cheaper at a slower rate than flash.

A manager looking at these trends would deduce that the lines will cross, and at that time, it will become cheaper to make personal computers with flash memory rather than hard drives. This smart manager would then work backwards from the anticipated crossover point, calculating the date at which the company needs to start designing HD-free computers, and the deadline dates for arranging the supply chains, and the ad campaigns for the new designs. Designs that *substitute* flash for hard drives.

Now, think about the speed at which the cost of Internet bandwidth is declining, and think too about advances in video compression. Together, they mean more video content can be streamed to viewers at a lower cost.

Do you remember getting Netflix movies at red kiosks? Or through the postal mail? Do you wonder why they didn't name the company KioskFlix? Or Mail-It-To-You-Flix? The name was Netflix from the start, because the founders anticipated price declines in Internet bandwidth, and anticipated techniques for squeezing videos into smaller files without much loss of image quality. They knew this would allow them to *substitute* streaming delivery for kiosk and postal mail delivery of movies.

Cordless electric drills occupy our workbenches today, and cordless electric shavers sit in our bathrooms. This is for the happy reason that batteries have become so much better. Measured in dollars per kilowatt-hour, battery prices have dropped twice as fast as experts predicted,[2] and the batteries last longer. This opens the market for additional kinds of products that don't need to be plugged in all the time.

---

[2]https://www.utilitydive.com/news/why-battery-storage-is-just-about-ready-to-take-off/407096/

On a bigger scale, electric cars, buses, and trucks now become feasible. This substitution of electric motors for gasoline engines reduces toxic emissions in cities. (But, we have to ask, is the battery manufacturing process—located, no doubt, in the countryside—polluting? Are we just displacing emissions from one location to another?) Beijing is a leader in market penetration of electric motor scooters. It's wonderful, but a pedestrian has to be alert because it's not possible to hear the scooters coming. Watch headlines for tallies of "cell phone zombies" getting clipped by silent scooters!

Gene sequencing. Flash memory. Batteries. Video compression. Like Moore's Law in integrated circuits, these other price–performance curves may have limits, may face their own "brick walls." We do not always know what these limits might be, or when they'll be reached, but we are sure they exist. Meanwhile, whatever your own product area, you might think about how you can use P–P curves to manage your company "by the numbers," as Intel and Netflix have done.

8.5   A Trend Is a Trend 'til It Bends

A trend may appear to be an orderly straight line,[3] until, bang! surprise! it goes off in another direction. The learning curve for polyvinyl chloride (PVC), ubiquitous in textbooks, is a perfect example of pioneer futurist Joe Martino's dictum that "a trend is a trend 'til it bends."

In Fig. 8.6, the horizontal axis is cumulative PVC production, and the vertical is the logarithm of the unit cost of production. (You may ignore the numbers on the axes.) This is the usual representation of a learning curve, more commonly known now as an experience curve.

The kink in the PVC learning curve could only have come from a manufacturing innovation. How to model this? I devised (Phillips 1999) (to Martino's delight) a nonparametric piecewise-linear regression for this purpose. The dark squares in the graph are the observed data points; the white squares are the fitted points from the regression.

Looking at the complete picture, it is surely obvious that two trends were at work. The trick is to determine, when observations start to depart from the straight line, whether the departure is an aberration (as we apparently see in the lower right of the graph), an error in measurement, or a transition to a new trend line. Equally obvious is that the PVC producers (and the stock analysts they communicate with) knew they had improved the manufacturing process. The rest of us had to figure it out from patterns of price reductions, or from data released by industry associations—hence the regression.

---

[3]Or a straight line when plotted on semi-log paper.

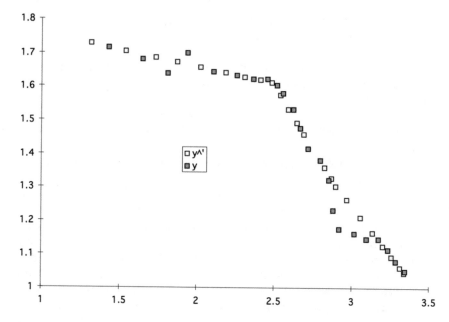

Fig. 8.6 An innovation in the manufacture of polyvinyl chloride, as evidenced by its kinked learning curve. Source: Author

8.6 Big Data, Big Models

Why are we enamored with "big data"? It's not the magic bullet of management or policy: it has made money for a few companies, but has backfired significantly in the arenas of national security (NSA surveillance scandals) and social media (OKCupid manipulating your emotions[4] without your informed consent; Facebook doing basically the same thing[5] and even worse, compromising USA election security by leaking data to Cambridge Analytics). E-commerce vendors serve you "customized recommendations" that couldn't interest you less.

Autonomous (no-driver) cars and trucks feature machine learning, the robotic drivers eating huge amounts of data on traffic situations. Tragically, an autonomous car killed a pedestrian in 2018. Industry chatter suggests at least a 5-year wait before the software is close to "ready for prime time," but even then not all the "corner cases" (the vendors' antiseptic term for accidents and fatal collisions) will be ironed out. Yet the vendors' perceived market window begins now, not 5 years hence.

---

[4]http://www.usatoday.com/story/news/nation-now/2014/07/29/okcupid-profile-experiments-online-dating/13308865/

[5]http://www.bbc.com/news/technology-28051930

At least in the short run, then, big data's potential tilts more toward pain than gain. Yet promising applications begin to appear:

1. Literature-related discovery, the searching of large bibliographic databases for obscure regimens for treating diseases, has turned up promising therapies for a number of common and rare maladies.
2. Data mining for technology fusion supports open innovation by searching patent databases for technologies that may profitably be combined to create innovative products.
3. Real-time big data for epidemiological control helps public health officials prevent or slow the spread of disease.
4. Big data analytics may shine in situations where the result just has to be pretty good, and not perfect. This usually means picking a target market segment for whom "pretty good" is good enough. For example, a Taiwan university hospital has analyzed 200,000 baby cries, creating a system that can tell from the cry, with 70–80% accuracy, whether the baby is hungry, poopy, or needing to be cuddled. An attentive mother can probably achieve 90+% accuracy, but the system is intended to alert nurses and aides in large hospital wards, whose efficiency is served by knowing whether to bring bottles or diapers when attending a crib.

These four promising areas have something in common. They do not allow computers, robots, or cars to make decisions. Rather, they deliver useful information back to humans. We might say they turn data into actionable information.

To be sure, managers may make poor or harmful use of the information, for example, by allowing it to be hacked, or by deliberately breaching confidentiality. Nonetheless, the four applications just listed reinforce our earlier conclusion that—at least in the short run, we must again say—*computer-aided* beats *computerized*. Expect to see more humans checking on and backing up AIs.

8.7   The Two Most Reliable Ways to Predict the Future

There are two very dependable keys to the future. The first is demographics, and the second is the Kondratieff wave (though the latter may be somewhat less precise).

Workforce sizes in the next 18 years are quite predictable, because everyone who will enter the job market within that time has already been born.

Population growth, as indicated by fecundity trends, is also dependable. We know not just the trend of fecundity, but we know *why* the trends are what they are. Increased affluence, expanded social welfare, and better medicine all lead to lower numbers of live births per fertile woman. This is because more infants survive to adulthood and parents need fewer children to support them in their old age.

We know further that cultural norms may lead to misuse of gender-selection technology. An example is Chinese parents' preference for boy babies. The result is an excess of men of marriageable age and a shortage of potential brides. The more peaceable resolution will involve increased migration, as Chinese men "import" wives from Southeast Asia and Tibet. The less peaceable option, which with luck will not happen, would involve renewed border incursions into Vietnam and other neighboring countries.

Every country's "population pyramid" powerfully foresees its future economic vitality, gender balance, and pension burden (Fig. 8.7 illustrates).

Blogger and physicist Johannes Koelman reminded me[6] that in 1963 we experienced a transition from a trend of increasing global population growth rates to decreasing rates. This growth rate reversal caused hyper-exponential global population growth to change to a trend toward a stabilizing global population.

The Russian economist N.D. Kondratieff (1892–1938) noticed 60-year cycles of faster and slower economic growth. Economic pressures in the 30-year "down-cycle" incentivize innovation. In the following up-cycle, riches are generated by exploiting these innovations. Further innovative activity languishes until the next down-cycle commences.

I have mentioned in earlier chapters that uncertainty completely dominates risk after 50–75 years, and that individual (or family) fortunes stop accumulating and start decumulating in 50–75 years. These phenomena are not proven to coincide with Kondratieff cycles, but they share the same periodicity, adding evidence that there are such things as long cycles.

There are signs suggesting the long Kondratieff cycle (K-wave) length may now be shrinking, from 50–60 years to 30–40 years, at least in the richer countries. Linstone (1999) noted that "the whole subject of [long] cycles remains a highly controversial one, with orthodox [technically]-oriented individuals, particularly economists and physicists, firmly opposed." Yet other distinguished economists, most notably Joseph Schumpeter and Simon Kuznets, embraced the idea of long cycles.[7]

8.8 I Make the Case for Replacing Business Executives with Robots[8]

This is no slur on the intellects of executives! In the transformed enterprises of the future, robots will take on more and more business decisions. (Humans will retain a smaller but still crucially important role.) Rather than e-business and m-business, we'll be saying r-business, for robotic business.

---

[6]http://www.science20.com/machines_organizations_and_us_sociotechnical_systems/crossing_over_how_use_and_misuse_fisherpry_curve

[7]Kuznets took a slightly different slant on the long waves; see Milanovic (2016).
https://voxeu.org/article/introducing-kuznets-waves-income-inequality.
See also Coccia (2018).

[8]The base option example stems from work done some years ago with Raj Srivastava. The principal-agent example is adapted from that of Katzman et al. (2009), which was in turn adapted from Wu (1993).

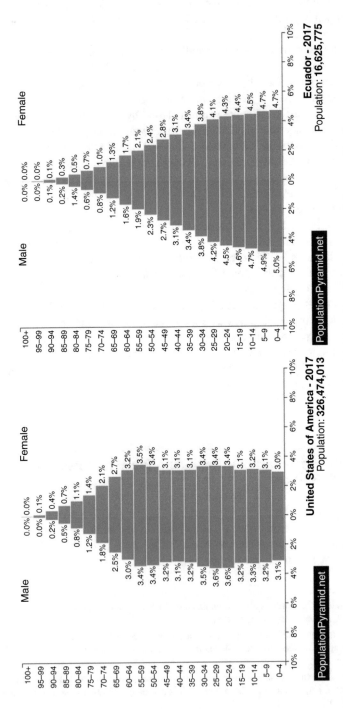

Fig. 8.7 Population pyramids: population by gender and age, 2017. Left, USA. Right, Ecuador. Source: populationpyramid.net

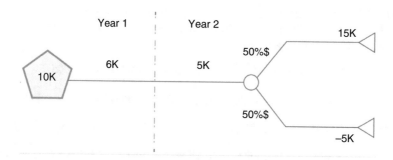

Fig. 8.8 Back to decision trees—this time to illustrate the principal-agent problem. Source: Author

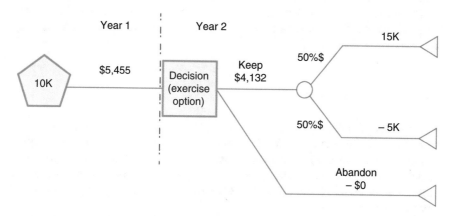

Fig. 8.9 We introduce a project discontinuance option. Source: Author

The argument involves "real options" and "agency theory." Explaining them is simple, though lengthy. So let's get started, again using a decision tree (Fig. 8.8), but a more complicated one than that of Chap. 2, as an illustration.

An opportunity requires Rineu Corporation to invest $10,000 now, with an assured first-year cash flow of $6000. The second-year cash flow is uncertain with a 50–50 chance of either a $15,000 gain or a $5000 loss.

The expected undiscounted gain in year 2 is 0.5 (15k–5k) = 5k. The cost of capital is 10%. The net present value (NPV) of the project is thus:

$$NPV = -\$10,000 + 6000/1.10 + 5000/(1.10)^2 = -\$413$$

The NPV is negative, indicating that Rineu should not make the investment.

Now, let us say that the project can be abandoned at the end of the first year if new information uncovered during this period suggests that the second-year payoff will not be favorable. Dropping the project at that time involves no salvage value or penalty. Figure 8.9 conveys these changes.

This NPV calculation didn't let us use the option of abandoning the project after we got the additional information. What if it did allow abandonment? Then we could redraw the decision tree, and write:

$$\text{NPV(ABANDON)} = -\$10,000 + 6000/1.10 = -4545$$

$$\text{NPV(KEEP)} = -\$10,000 + 6000/1.10 + 15,000/(1.10)^2 = 7852$$

$$\text{NPV(OPTION)} = 0.5 * \text{NPV(KEEP)} + 0.5 * \text{NPV(ABANDON)} = \$1600.$$

The now-positive NPV implies that we should make the investment, assuming (1) no alternative projects show higher NPVs, and (2) we can trust a manager to make the right follow-up decision at the end of year 1.

The example confirms what is well known, namely that traditional discounted cash flow analyses can understate the attractiveness of new ventures that have highly uncertain initial returns.

8.8.1 "Expectations"

Now an important digression about decision trees—in fact, about all decisions based on mathematical probabilities. Most mathematicians were math majors, not English majors. They're good with equations, not always so good with words. "Expected value" is an example. Mathematicians' equations, especially equations as simple as those we have used in the decision tree, cannot describe the complex psychology of what I do or don't expect.

In fact, if I use the tree to make a single, one-shot decision,[9] there's every possibility things won't turn out as I expect. This is because the tree uses probabilities. No matter how high the probability that an event will happen, even 99%, there's the small chance that it will not happen.

Do you have a right to "expect" that the event will occur? Surely, if that's what you want to do. (You might be disappointed.) Much depends on whether you're an optimist or a pessimist, what kind of mood you're in today, and so on. The math doesn't know those things.

The mathematicians should have stuck with the term "mean value," which has fewer psychological overtones.

However, if you make the same decision under the same circumstances many times, you can in truth expect the "expected value" to eventuate, on average. This is the "frequentist view" of probability—to which I subscribe. There are other views of probability, called subjectivist or Bayesian, in which words like "feelings" and

---

[9]Using the expected value criterion. There are other criteria for pruning the decision tree, for example, the "minimax regret" criterion. We won't examine those here, as they're explained in almost any decision-making textbook, and because I hope you will aim for a positive future. Not like those grim "minimax regret" folks who try to minimize their expected maximum disappointment.

"knowledge" are freely used. I for one reject these because I think it's absurd to believe that these simple probability formulas say anything about human psychology.[10]

So again, the point is that making a one-time decision based on mean values can be the best you can do, but is not totally reliable. It becomes reliable when you make repeated decisions in identical circumstances. Identical circumstances may arise in stock trading, for example, but they rarely occur in organizational management.

8.8.2   Real Options

A "real option" means treating a contingent operational decision using the same reasoning one would use for a financial option. Operational projects take on a higher initial valuation when one can count on a decision maker making the best use of information that is expected to arrive at a later time.

Exercising a financial (stock market) option is easy: your broker will remind you when the expiration date or strike price is reached. As you can imagine, though, the communication and control (C^2) issues are more complex when a corporation faces a real option that involves many employees, departments, or divisions.

When choices and "chance branches" are discrete and few, the "decision tree" analysis used before is a sufficient solution technique for a real option.

However, many operational decisions are characterized by continuous probability distributions, and by continuous-choice outcomes (e.g., "Invest $x in the project," where x is any non-negative number). In these instances, the math is much hairier, sometimes using the Black–Scholes model,[11] integral equations, or other scary creatures.

Our discrete math example, though, suffices for the r-business argument.[12]

8.9   If Options Are So Great, Why Is NPV Still Widely Used?

Acquiring a fleet of trucks (for example) has one value to the acquiring company if later decisions about maintenance are made in a certain way, for example, if the oil is changed at prescribed intervals. The fleet shows a different lifetime value if the oil is changed less frequently or not at all!

---

[10]In certain well-defined situations, the entropy of a probability function can measure knowledge. This was discovered by the famous Claude Shannon and set forth in his 1948 *The Mathematical Theory of Communication,* and later applied (in my own doctoral dissertation) to a kind of forecasting.

[11]http://en.wikipedia.org/wiki/Real_options_valuation

[12]And involves nothing more complicated than Bayes' Theorem. Did I say "nothing more complicated?" Well, Bayes' theorem is very simple mathematically, but terribly contentious in its interpretation.

If the return on fleet acquisition is allowed to depend on reliable mechanics changing the oil regularly, why are other kinds of project selections made without depending on reliable executives to make the subsequent decisions that maximize the project's value? Some years back, I surveyed high-tech executives on this question. Here are their replies (in descending order of reported importance):

Survey Scores: "Reasons Why Your Company Does Not Use the 'Real Options' Method of Valuing Projects"?

- Perfect information for project evaluations at future points is rarely available (or difficult to obtain).
- Operations executives do not like to discontinue their own projects at a future point of evaluation.
- All possible "options" cannot be anticipated.
- It's more convenient to obtain complete project funding now, rather than complete for partial funding with other projects in the future.
- Conservative decision making avoids choosing "options" or alternatives that involves large downside risks.
- The company's entry and exit barriers for projects will not permit project expansion or discontinuance based on options reasoning.
- Employee turnover and transfers make future project expansion/discontinuance difficult.
- Evaluating a project at each future decision point may incur higher costs.
- Project valuation is generally performed by financial rather than operations executives.
- Options are more complex than ROI/NPV.

Personnel turnover means the person who is charged with making the later decision may no longer be with the firm at the time the decision is due. If records are not complete, his/her successor may not be able to step in effectively. Then too, as implied by the survey results, the financial exec making the initial valuation may not trust the operations exec—who works in a different department—to make the right decision at the right time.

It may strike you as absurd that a company trusts its truck maintenance foreman more than it trusts an executive who makes 100x the foreman's salary. We now zero in on the trust issue, in the form of the "agency problem," a.k.a. the principal-agent problem.

8.9.1 The Principal-Agent Problem

Referring to the abovementioned decision tree, suppose the decision to exercise the option is left to a risk-averse, loss-averse manager. The manager is awarded a bonus of 5% of profits. As regards the project in question, our manager's private calculation adds a risk premium to the discount rate (cost of capital), making it 12% instead of 10%. His "utility" for money is concave, meaning that his second million dollars add less pleasure to his life than the first million did.

We can hammer together a quick-and-dirty utility function for him (tailored to the example, but, as the mathematicians like to say, "without loss of generality") that looks like this:

$$\text{Utility} = -200 + \sqrt{(\text{income} + 2000)}$$

The executive's utility is the square root of the sum of his income and 2000, less 200 dollars. (Square root of x is a fairly elementary *concave* function of x.) He has already nailed the bonus on 2500 in profits earlier this year. He sees the current opportunity in this way:

$$\text{NPV(ABANDON)} = \sqrt{[0.05\,(2500 - \$10{,}000 + 6000/1.12) + 2000]} - 200$$
$$= -\$156$$

$$\text{NPV(KEEP)} = \sqrt{[0.05\,(2500 - \$10{,}000 + 6000/1.12 + 5000/(1.12)2) + 2000]} - 200$$
$$= -\$149$$

$$\text{NPV(OPTION)} = 0.5 * \text{NPV(KEEP)} + 0.5 * \text{NPV(ABANDON)} = -\$152$$

Let's suppose that Rineu Corporation (as its name cleverly implies) is risk neutral (or more generally that the controlling shareholders—the "principals—have a risk profile that differs from that of their agent, the manager). If risk neutral, the shareholders would expect a "yes" decision on this project because of the positive NPV of the option. The manager's motives and personality, however, drive him to a "no" decision that's in line with his personal utility but at odds with the interests of the shareholders. This is the classic "principal-agent problem," as it manifests itself in the real options situation.

8.10   What Can Be Done?

Some companies fiddle the bonus formula, trying to align the interests of owners and agents. This rarely works, because the manager-agents' utility function is obscure even to the managers themselves, and changes with their moods in any event.

Other companies say, heck with it, we'll just go with straight NPV. They're leaving money on the table by taking this route.

In some industries, companies do use real options on a routine basis. One such is the package delivery business. Pick-up and delivery orders in each 9-digit zip code (postal code) arrive according to by-now well-tested probability distributions. The value of mobilizing x additional trucks and y more driver shifts (or diverting a particular truck from its current route) depends on the pattern of orders that materialize subsequent to that decision. Drivers are, of course, in constant communication with dispatchers. A perfect situation for using real options.

Why is it perfect? The added profit margin from taking best advantage of truck and driver availability is significant. The decision situations happen routinely, indeed several times per hour. The decisions are of a uniform kind. C^2 problems are minimized, because everything happens within the database that drives, no pun intended, the whole business.

Who makes the decisions? A robot. Not Robbie the Robot, of course, but an algorithm. The robot knows the truck maintenance schedules, knows the neighborhood maps, and knows not to schedule triple shifts. It makes best dispatch decisions based on these and other constraints, and on evolving patterns of pick-up orders.

With a tip o' the hat to Tom and Ray Magliozzi's Russian chauffeur, we'll call this robot Pikop Andropov. There is no principal-agent problem with Pikop. Pikop has no utility function of his own; he can be easily tuned to the same risk profile as the shareholders', taking full advantage of operational options in a trusted manner. Pikop's control limits and alarm whistles bring a human operator running if and when truly unusual situations arise.

8.11 The Future

As enterprise computing systems mature (Oracle, SAP, and the like are all still pretty messy), the C^2 costs and risks of real options will decline. The "big data" and "business analytics" movements will make more industries' information flows look like those of the package delivery industry.

DHL and Federal Express, already for all intents and purposes e-businesses, have become r-businesses. They leverage real options, and have got rid of routine principal-agent problems. Human executives are there for strategizing and for HR functions. Tactically, they are responsible only for dealing with exceptions and emergencies.

My students in Korea, where I taught for some years, tell me Korean companies already operate in this manner, but in a slow, human-intensive, analog way. Rules for most situations are found in "manuals." Human managers react to situations by looking for procedures in the manuals. They are permitted to call on higher-ups only when no manual addresses the situation at hand.

So, I asked them somewhat facetiously, if managers are not allowed to make decisions flexibly... why are you studying for a Master's degree in management? Naturally the question flustered them, and I got the answer I expected: none.

The implication for management educators is clear, however. Human managers and executives of the future need to be taught not routine decision making (which is what we've taught in the past), but emergency management: how to prevent, remediate, and minimize the impacts of crises and disasters. And how to creatively profit from exceptions to the norm. Robots will handle the rest.

Key Takeaways

- Though most of today's complex problems do not yield to strictly mathematical solutions, some mathematical curves—S-curves, price–performance curves, substitution curves, and others—remain central to the forecasting endeavor.
- So far, demographics and the Kondratieff wave are the two most reliable ways to predict the future.
- As computing systems in corporations become more organized, big data and business analytics will enable machines to deal with routine tasks effectively, leaving only exceptional situations for humans to handle. Flexible decision making in moments of crisis will be a critical skill for future managers to master.

References

Coccia M (2018) A theory of the general causes of long waves: war, general purpose technologies, and economic change. Technol Forecast Soc Chang 128:287–295

Katzman BE, Verhoeven P, Baker HM (2009) Decision analysis and the principal–agent problem. Decis Sci J Innov Educ 7(1):51–57

Linstone HA (1999) Complexity science: implications for forecasting. Technol Forecast Soc Chang 62(1–2):79–90

McKelvey JP (1985) Science and technology: the driven and the driver. Technol Rev 88:38–74

Milanovic B (2016, February 24) Introducing Kuznets waves: how income inequality waxes and wanes over the very long run. Vox EU

Phillips F (1999) A method for detecting a shift in a trend. PICMET '99, Proceedings of the Portland International Conference on Management of Engineering & Technology, Portland State University, Portland, Oregon

Phillips F (2007) On S-curves and tipping points. Technol Forecast Soc Chang 74(6):715–730

Phillips F, George Hwang G, Limprayoon P (2016) Inflection points and industry change: was Andy Grove right after all? Int J Technol Manag Growing Econ 7(1):7–26

Wu G (1993) Decision analysis. Harvard Business School Case 9-894-004, rev. December 4, 1997

Imagination and the Future

9

> *Those who cannot change their minds cannot change anything.*
>
> —George Bernard Shaw

This chapter introduces the idea of the techno-management imagination (which we'll abbreviate as TMI), and its role in apprehending the future. TMI is the process by which, in times of organizational crisis, managers think not just out of the box, but out of the very reality in which the box resides, thus accessing technological solutions that were unknown and/or unimagined. The chapter addresses the interaction of technology not just with the artistic and creative imagination as conventionally conceived, but with the perception of ideas and methods that are outside the boundaries of the manager's cultural milieu.

Because these ideas and methods are disallowed by the culture, they are commonly labeled terra incognita, superstition, wizardry, secret sauce, spirituality, or magic. They are thereby placed in opposition to "knowledge," which is thoroughly embedded in the home culture. One purpose of this chapter is to extend scientific discussion toward the treatment of the imaginative unknown. Another purpose is to highlight the important role myth plays (often without our conscious awareness) in shaping the technological futures we strive for.

TMI is being exercised more freely and more widely now than in much of the twentieth century, though perhaps not as freely as earlier in history. This is occasioned by rapid changes in the socio-technical environment, especially information and communications technology enabling decentralized, alliance-based, and entrepreneurial initiatives. The globalization of technology industries is also a factor, as projects become cross-cultural, and affordable information and communication technology (ICT) democratizes technological initiatives. Managers and scientists should take care to balance the usefulness and urgency of embracing the unknown against the

This chapter consists of excerpts from F. Phillips, "Technology and the Management Imagination." *Pragmatics & Cognition,* 13:3, 533–565, 2005, reprinted by permission of the publisher.

danger of anti-intellectualism and other side effects that can endanger the foundations of technological society. Science can assist this evolving balance by expanding its boundaries to build on newly observed managerial and technological phenomena.

9.1   Socially Constructed Reality

According to an oft-told story, when his exploration vessel first anchored off the reef of Tahiti, Captain Cook watched for a reaction from the natives. There was no reaction. Later conversations revealed that Tahitians on the beach did not see the ship, though it was in plain sight and quite large. How could they fail to see it? The answer has to do with mental categories, denial, and "consensus trance."

Hundreds of years of experience had confirmed that anything on the surface of the ocean was flotsam, a canoe, or a basking whale. Any object not fitting one of those categories could not exist. If a few Tahitians did register something out of the ordinary, they may have been embarrassed to mention it to the majority who showed no signs of noticing. The process by which we actively and tacitly persuade each other to see and believe only a certain subset of the whole is called *consensus trance*, or the *social construction of reality*.

Foreshadowed in the poetry of William Blake, these ideas were first subject to scientific inquiry by Aldous Huxley (1963). Searle (1997) argued against both materialism (all reality solely objective) and solipsism (i.e., that reality is solely subjective), in favor of the view that some facts are independent of human observers, and some ("intersubjective" facts) require human agreement. The controversial anthropologist Carlos Castañeda (1970) put these views into somewhat more accessible language. The cited works imply: (1) intersubjectivity is essentially what used to be called "mob psychology" and (2) it applies to a much wider range of experience than crowds and violence; in fact, the intersubjective portion of our daily experience and deepest assumptions is far larger than we might expect.

Does socially constructed reality prevent modern business people from seeing "Cook ships"? Yes, it does. When chaos theory emerged from biology and physics, it became clear that "deterministic chaos" must also be present in market research data. N.W. Kim and I showed (Phillips and Kim 1996) how marketing professionals had inadvertently constructed a culture that prevented them from seeing it, despite that taking advantage of these data fluctuations could lead to increased profitability.

The following provocative passage supports the thesis that the root of sociotechnical magic is a *shared, implicit, and incorrect view that certain things are impossible.*

> Management should never underestimate the constraining effects of technical folklore and deep-rooted prejudices shaped by the accumulated successes and failures of generations of technical products. These attitudes breed barriers so fundamental that they are unrecognizable except to outsiders. Because of such barriers, many companies repeatedly market products that never overcome fundamental flaws and never achieve distinction. Such organizations believe, instinctively and without debate, that certain directions are closed to them, even as other organizations proceed in exactly those directions with success. (Rudolph and Lee 1990)

Corporate cultures that repress initiative and lack open lines of communication cannot benefit from the full variety of skills and interests of their constituents. However, what distinguishes TMI from the more ordinary managerial tasks of respecting diversity and leveraging the varied skills and opinions of a diverse employee base are questions of *seeing, denial,* and *teaching.*

- When a "ship of Captain Cook"—something that will change the firm's future irrevocably—appears on the company's horizon, will the company see it? The CIA did not foresee the fall of the Berlin wall. It's not just that they didn't foresee the date; they didn't foresee the *possibility.*
- In pursuit of a cohesive company culture, does the company channel employees' perceptions too narrowly, preventing them from seeing Cook ships? S.I. Hayakawa, the linguistics scholar who was president of San Francisco State University, noted, "If you see in any given situation only what everybody else can see, you [are] so much a representative of your culture that you are a victim of it."
- The CIA did foresee the possibility of the 9/11 tragedy, but—whether due to communication gaps, disbelief, or different priorities—there was a disconnect between the Bush administration and the agency, and hindsight tells us certain actions should have been taken but were not.
- Some Asian languages place little emphasis on the difference between the phonemes "l" and "r." Indeed, Western linguistic science considers them separate phonemes, and Japanese linguistic science (for instance) does not. At the practical level, never bothering to pronounce this difference while growing up, adult speakers of those Asian languages are rarely able to *hear* the difference between els and ars. They consider it mysterious that Westerners can conversationally distinguish the meanings of "lead" and "read." Developmental pathways that allow us to hear the difference do not mature, if we live in a culture that does not place emphasis on the distinction. Of course, very young Asians exposed to Western languages have no difficulty articulating els and ars.

9.2   Three Flavors of Socio-technical Magic

A behavior and its outcome may be viewed as magic in any of three circumstances. We'll call them intercultural magic, ancient magic, and innovative magic.

9.2.1   Intercultural Magic

Isolated human cultures of the past had few shared concepts. Their differing living environments led to different senses of what was important. This, in turn, focused their consciousness in ways that led to seeing certain things and not seeing other things. This kind of focus was a survival mechanism.

Thus, a member of Culture X routinely did things that a member of Culture Y (think Navajo and Hopi, or perhaps European crusaders and Middle-Eastern Arabs) would have thought impossible, or at least incomprehensible.

When Culture X and Culture Y collided, because of expanding settlements or chance encounters of hunting parties, each side called the other's odd behavior, and its results, "magic." Chieftains and shamans wishing to maintain the integrity of their own societies would label the other culture's behavior "evil magic," and forbid their own folk to practice it. Today, we are quick to label this attitude as inimical to innovation. However, the ancient tension between deviant behavior and continuity/conservatism of culture is still with us.

It looks like magic when one culture achieves a behavior/outcome pair that is beyond the ordinary horizons of another culture. A member of a traditional subsistence whaling/sealing culture would, no doubt, find it astonishing that modern Americans teach orcas to fling human swimmers high into the air, and let our children place fish in the "killer whale's" mouth.[1] These behaviors are not only possible, they are fun and fairly safe. We may conclude, therefore, that no influential person in the traditional village (where whales were *of course* either quarry or rival seal predators) ever tried to do these things—or, if they thought of it, had no time to attempt it—and so it never occurred to the other villagers that it could be done at all.

In turn, we are astonished by the ways and the reasons Amazonian shamans utilize snakes, frogs, insects, and plants. Enterprising pharmaceutical firms are embracing this magic (delicately renamed "ethno-pharmacology") as a cost-effective alternative to discovering disease-fighting drugs in the laboratory. In highly competitive and regulated pharma markets, such drugs can save companies as well as people.

We are also surprised that traditional Asian physicians can diagnose illnesses by taking a patient's pulse and smelling the patient's breath or urine. Western M.D. s confirm these techniques work for a wide range of ailments, but understand that in industrialized nations where medical school and malpractice insurance are expensive, and automated blood diagnostics cheap, there is little point in training doctors in the traditional Asian methods. Early in this century, however, a US university, exercising excellent TMI, developed a sensor that "smells" abnormal acetone levels in a patient's breath, diagnosing early-stage diabetes more reliably than most other mechanisms.

Corporate cultures seem susceptible to the same effect. When Daimler and Chrysler merged, for instance, executives in each company viewed the others' practices and asked incredulously, "How can they expect *that* to work?" (Executive incentives were one specific focus of skepticism.) The fact was that in each original company, the questioned practice had worked well and long. In the merged company, it could not, and performance of the merged entity bent under the weight of perceptions of bad magic.

---

[1]This example is adapted from Salzman (1991).

9.2.2   Ancient Magic

Second, a behavior/outcome pair may constitute social magic when it had been a part of a culture, but is no longer needed for the society's survival. The practice may be maintained, by a few, as a cult, a hobby (amateur rocketry would be pertinent to the example below), a specialized niche market, or for entertainment. Maintaining these practices can require extraordinary skill and dedication. In spite of this, or perhaps because of it, practitioners are socially isolated. Shamans, wizards, and artists maintain the integrity of their work by deliberately becoming loners, or even hermits, in order to isolate themselves from social hypnosis.

Where would ancient magic be useful?

- After a generation of space shuttles, NASA has forgotten how to launch high-payload, high-orbit spacecraft. If this knowledge can be recovered, it will be because of carefully archived notes and the capture, perhaps via expert systems, of retired engineers' recollections.
- Finishing violins in the manner of Stradivarius would be of great artistic and economic value today, but the knowledge is lost.
- The first known man-made underwater tunnel joined the two fortified halves of the ancient city of Babylon, in 2160 BC. Nearly 4000 years would pass before the next known tunnel under a river, a passage to transport carts under the Thames.[2]

9.2.3   Innovative Magic

A third category, "innovative magic," is characterized by very advanced science—often theory untested by experiment and for which a practical application has not yet been imagined—combined with audacious engineering, mobilized to save an organization from extreme crisis or competitive pressure. In 1980, shrinking state budgets and the Bayh-Dole Act created a new environment for US higher education. Universities responded by searching their laboratories for commercializable advances that had remained under the administration's radar. A few universities reaped substantial license income from this new strategy.

9.3   How Companies Use TMI Today?

Intel is a leading exponent of the "copy exactly" philosophy, which dictates that the first fab to successfully manufacture a new generation of chips should be copied exactly for subsequent facilities, in order to minimize the chance of large downside variation in wafer yield. Another semiconductor firm reportedly built an initial successful fab that, due to a plumbing error, featured a urinal installed five feet

---

[2]Popular Science (no byline). 1995. "The facts." *Popular Science:* December, 102.

above the floor. This company built its second fab—and deliberately placed a urinal in the same inaccessible spot!

Both companies, under intense pressure to recover huge fab construction costs within a short time window, substituted "copy exactly" for a complete scientific understanding of the new chip's manufacturing process. The companies say "copy exactly" demonstrates prudence; others call it superstition. Differences in personnel, training, local culture, management style, etc. surely result in more yield variation than the placement of the pissoir. In any case, "copy exactly" exemplifies reaching toward the unknown for needed answers at a critical time in a company's history.

In fact, Intel places very little value on *understanding* the semiconductor process.[3] More engineers than scientists staff Intel's research labs. Engineers "understand" things, of course, and are systematic in their work. But Intel's research staffing shows the company's greater interest in seeing what works, and how to replicate it, than in developing new theory. Intel's need for speed makes the search for theory uneconomical, at least when the theory may be valid for only one chip generation or less.

Referring to any scientific or professional subject, Einstein said, "If you can't explain it to a ten-year-old, you probably don't really understand it yourself." Just as well that Intel places small value on understanding; to speak of "understanding" a microprocessor with billions of transistors, which was itself designed in large part by intelligent software, is almost laughable. It is possible only in terms of its testing protocol: if you put that set of signals into a good chip, these numbers should come out. Bad chips, defined and identified as such when these numbers do not come out, are simply discarded. (It may be possible to know how the processor changes a particular input into a certain output, but no one can maintain a mental model of how it transforms myriads of ensembles of inputs.)

Noting the "fluid sense of perception [and] willingness to tinker with cognitive structures" of educated users of hallucinogens, Intel and other major corporations give employees ample advance warning for the urine tests they are required to undergo (Kim 1991). The desire for creativity in hypercompetitive conditions, he notes, leads to the high-tech industry's "no-sweat attitude toward chemical recreation."

Firms access magic/TMI by using:

- *Environmental scanning.* Looking outside the firm for new behaviors and technologies.
- *Alliances.* Giant pharmaceutical companies ally with start-ups and universities for drug discovery. In World War II, the British tapped Polish code-breakers, and the Americans utilized Navajo code-talkers.
- *Wizards.* Intel, Xerox, and other technology firms have Fellows programs. Fellows are allowed wide latitude in their choice of projects, and have no corporate management duties. They enjoy the perks of the wizards of myth:

---

[3]Yeh and Yeh (2004).

autonomy, solitude, and recognition without the responsibility of political authority.[4]

Wizards are found not only in "wizard farms" such as Fellows programs; they may dwell in unexpected places, either within the firm or without.

- The costs of developing a new airframe are so high that Boeing bets the company on every new design. A modern jumbo jet has three million parts. Boeing's 777 development was speeded by a "fly-through" virtual reality system for assembly simulation—a system invented by a physician who hoped to use the technology for surgical training.
- SAP looks for wizards throughout the firm (Muehlhausen 2004). IBM and Siemens are committed to looking for innovative behavior in "[internal] research... the venture capital community... universities, research institutes, other companies... and the whole environment."

We lose magic when our circumstances evolve. Indeed, hindsight makes what was ordinary before the change magical and legendary after the change. When do we *need* magic? *When circumstances change radically again*, as they are sure to do. Then, old behaviors no longer ensure survival. We have to look to our own old practices, borrow practices from foreign cultures, or invent brand new practices in order to survive.

Using "magic" as a synonym for "something to be resisted," Stuart Cohen, CEO of Open Systems Development Laboratory, remarked, "People thought Linux was magic, that it came mysteriously out of nowhere." When the time comes when magic will save a company, it must have mechanisms in place to help employees see, accept, and perform the magic. Just as the painter Vermeer taught us to *see* the play of morning light through a window, companies have used artists to advance these mechanisms. Shell Oil hired storytellers to turn the company's alternative-futures scenarios into compelling, logical narratives. For the past 10 years, Xerox PARC has had a prominent artist-in-residence program. The MIT Media Laboratory has emerged as the leading proponent of artistic participation in the research process. Other firms have actually used stage magicians in their creativity seminars.

---

[4]"Intel Senior Fellows represent the company's most exceptional technical professionals," said Craig Barrett, then Intel's chief executive officer. "Their contributions have helped Intel maintain a position at the forefront of technical innovation and to continuously deliver cutting edge technologies to the marketplace." The Intel Fellow program began in 1980, and in 2002 the company further recognized "the most senior and influential members by creating the new role of Senior Fellow." Employing a total of 65,000 employees, Intel in 2002 had about 50 Fellows and about a dozen Senior Fellows.

9.4   Historic Interplay of Technology and Magic

Can there be a philosophical/historical justification for TMI, for technology managers' forays into the unknown? Plato would have said yes; he danced back and forth between rationality and mysticism, and believed the two could not easily be separated. Pascal remarked, "There are but two equally dangerous extremes—to shut reason out, and to let nothing else in."

In an excellent modern presentation that I cited in Chap. 6, E. Davis (1998) explores the interaction of technological progress and the transcendent imagination.[5] The following elaboration of 12 of Davis' implications ignores, for present purposes, his distinctions among "soul," "spirit," and "magic":

1. *Technologies, real or imagined, are at the very heart of some mythic traditions.* The Masons' view of themselves as the successors of the architect of Solomon's temple is the perfect example of the Gnostic tradition, equating effective knowledge with the highest pursuit of the spirit. Hermes, messenger of the Greek gods, took a lower road (the Greek *techne* meaning both technique and "trickiness"). That the Masons pursue "hermetic" knowledge illustrates the long-standing link between technology and transcendental imagination.

2. *Technology enables the realization of magical potentials.* Human flight, influence at a distance, strange hybrid animals, etc., are now of course a reality. Modern communications, weaponry, and the World Wide Web give new voice to ancient and obscure beliefs.

3. *Mythical and magical urges help map the possibilities of new technologies.* Example: the "avatars" that represent participants with different levels of privilege and power in computer games and in 2-D and 3-D virtual meeting spaces.

4. *We turn technology to serve the mythic imagination.* Example: Quest games like Myst. Perhaps more extreme are astrological software, I-Ching CD-ROMs, and Tarot hypercard stacks.

5. *Because new technologies are wonderful, they make imagination and magic more attractive to the uneducated.* Heron (10–70 AD) built machines (including one that appeared to turn water into wine) the purpose of which was to mystify, impress, and baffle. They "paradoxically eroded the cultural authority of the very rational know-how that stimulated their design in the first place," as Davis puts it.

6. *Technology can substitute for myth, thus serving as a life jacket for the spiritually sinking,* even if the technology is untested, merely metaphorical, or outright

---

[5] As did Nikolai Gogol in his 1836 short story "The Nose": "At that time everyone's minds were tuned to the extraordinary. Just a short time before, the public had been amused by experiments in magnetism.... Someone said that the nose was presently in Junker's store, and... a huge crowd gathered.... A respectable-looking entrepreneur with sideburns, who sold various cookies next to the theater entrance, made excellent, sturdy wooden benches and invited the curious to stand on them at the price of eighty kopeks a head." As for the possibility of magical occurrences, Gogol adds, "Say what you may, but such events do happen—rarely, but they do."

fake. Employing the command-and-control language of the then new science of cybernetics, L. Ron Hubbard's cult was an attractive beacon for people who felt they had little control over their own lives. Another way of saying this is that mythic yearnings enable behavioral technologies; Davis cites "propaganda, advertising, and mass media, those modern machineries of perceptual manipulation that often explicitly deploy the rhetoric of enchantment."

7. *By the same token, wonderful new technologies force us positively to reframe our myths.* David Learner and I described (Learner and Phillips 1993) how Van Leeuwenhoek's microscope opened new vistas of scientific data, for example, microorganisms. Davis adds that these vistas "evoked a sense of wonder and mystery, forcing us to reconfigure the limits of ourselves and to shape the human meaning, if any, of the new cosmological spaces we found ourselves reflected in."

8. Moreover, *each major new basic technology offers new language and metaphors that enhance our persistent myths—and thereby prescribe applications for the technology.* The invention of writing enabled the "religions of the book," Judaism, Islam, and Christianity—and methods of preserving the holy words. Electricity let us "sing the body electric." Reconceptualizing the soul as an electric field, medicine progressed from rigorous galvanic experiments on frogs to bizarre attempts to measure and influence electric phenomena in the human body.

9. *Unfamiliar technology colonizes the mythic imagination.* In Chap. 6 we noted that it may do this in the form of demons, etc.: "Alien abduction hallucinations constitute an example; they are anxiety dreams about our own mutation."

10. Furthermore, *technology colonizes and animates the extensive universe itself.* Not just in the clutter of machines surrounding us. The creation and study of computers leads us to reconceptualize the universe as an information processor. As man-made computers and computer-based simulations become more powerful, "the universal machine becomes a machine that builds universes." Whether a direct human intention or a matter of complexity theory and AI, researchers have put forth compelling evidence that in the 1990s "the networked noosphere began an irreversible process of self-organization" (Both quotations are from Davis.)

11. *As hyperlinked technological creation grows complex* (temporarily?) beyond what rational-analytical epistemology can comprehend, we fall back on tribal methods of knowing: "cultivation... pacts, lore, and guiding intuitions," again in Davis' words.

12. And thus, *technology is the root of modern terror.* Marshall McLuhan said that in highly connected societies, terror is the natural mode of existence, because everything affects everything. Early subsistence-economy societies were tightly connected with their environments because there were no surplus resources to cushion the impact of drought or flood. Today, it is technology that makes globalization possible, tying currencies, product distribution, and politics together worldwide.

Writing disseminated the religions of the book. Printing boosted the efficiency of dissemination, as Gutenberg knew it would. Religion was the cause and the effect of printing. Rational technology created LSD, which flings the user into a nonrational world and bounces back to better understanding of schizophrenia and new technologies for psychedelic music. Thus, technologies refine, give expression to, and realize our magical imagination—which forces technology in new directions and causes the cycle to continue. While giving full due to utilitarian uses of technology, Davis echoes philosophers Heidegger and Ellul in maintaining that our primeval myth stories and images are the primary driver of our technology choices, and the primary role of technology is to create new ways of knowing and being.

After a false start with a "sphere of knowledge" metaphor, Phi Beta Kappa journal writer J. Churchill (2004) recovers: "That metaphor of a sphere is misleading, as if we could glance about and see the work of our confreres. We are more like miners, burrowing away from our once-shared galleries, down shafts. . . convergent, parallel, divergent" (Churchill 2004). The mining metaphor, not a bad one, supports Davis' thrust in point #11. Speech is suited for narrative and negotiation, Churchill says, but the Internet lends itself well to hypertext; it is no surprise, therefore, that we cannot comprehend the universe of online knowledge. Churchill notes that the problem has been evident at least since the 1940s, when Bertrand Russell found himself unable to state his philosophical positions in ordinary English.

Thus, there are ample philosophical basis and historical precedents for cautiously exercising the techno-management imagination. Indeed, economic historian Henri Pirenne noted that even in the Middle Ages "capitalists. . . incapable of adapting to the conditions that demand needs hitherto unknown and requiring methods hitherto unused. . . In their place arise new men, bold, entrepreneurial, who allow themselves audaciously to be driven by the wind. . ." (In Hall 1998)

9.5 Socio-Technical Magic: Risks

Inhabiting the social fringe, the magician risks developing amazing skills that are relevant to nothing and no one. He or she must balance the need to hone the chosen skills against the possibility of losing all communication with society. The wizard Thomas Edison was eccentric but tolerated. His contemporary and rival Nikola Tesla, probably the greater intellect of the two, could not or would not come in from the cold regions of the far social fringe.

Other risky aspects of magic:

- Younger people who are still forming a sense of their character and limitations are also at risk if they attempt the wizard role. See Goethe's *The Sorcerer's Apprentice* for support of this point.
- In my 2001 book *The Conscious Manager*, I described a variety of supra-normal phenomena, and noted that fascination with them can lead to estrangement from

one's peers, as well as the other perils of idle distraction. I recommended focusing on the everyday affairs of one's business, training the consciousness (perhaps through meditative practice), and having confidence that one will thereby develop TMI and see when magic is needed and, then, where it may be found.

- This same kind of focused practice can help a manager distinguish between intuition and egotism. Listening to intuition, the manager's crisis decisions may be right more often than not. If on the other hand he or she egotistically shoots from the hip because "If I say it, it's right," then the decision will be wrong as often as right.

- In his pioneering work in mathematical decision theory, Pascal (whom I quoted earlier, and who was also an advocate of meditative practice) essayed an analysis of the utilitarian value of religious faith. This analysis is now universally regarded as fallacious. He was trying prematurely to integrate knowledge from his rational inquiries with knowledge from his mystical-philosophical investigations.

Companies risk tarring their reputations if visionaries and wizards are not managed wisely. Yet companies fail due to lack of imagination. Resources are misallocated because they do not understand the socio-technical forces which are in play.

One venture capital investor speaks of "magic" technologies, noting that VCs invest in entrepreneurs whose technology the VCs don't understand. If the statistics are any indication (only 1 in 60 VC-reviewed business plans leads to an IPO), this is still a badly understood process.

In a larger sense, our civilization may lose the capacity (as the animated world reasserts itself and as machines themselves create new industrial designs and make important decisions) to distinguish between the realm of human decision, design, and investigation on the one hand, and the simple wonder (magic) of our new environment on the other.

This last point is evident in declining enrollments in science and engineering programs and in the waning level of public technological initiative, in particular for space exploration, that Ray Bradbury mourned in a *Wall Street Journal* op-ed (Bradbury 2004). Perhaps in part because we feel less need than we did in the 1950s to seek new wonders elsewhere, we deliberately risk the highest stakes of all—keeping human civilization confined to one lonely, vulnerable planet.

9.6 Further Comments on TMI

An influential *Harvard Business Review* article (Nevens et al. 1990) correctly urged companies to prepare to receive and implement innovation, by building cross-functional skills and communication and by building familiarity with a wide range of technologies. However, the authors' emphasis on goal setting, benchmarking, and performance measurement implied a buttoned-down, systematic approach to innovation that was discomfiting. Not only does the latter view seem at odds with millennia of heroic-mythic tradition (viz., *The Lord of the Rings*), but one suspects that companies, when asked, exaggerate the systematic nature of their innovation processes. After all, it would hardly do to tell stock analysts that the firm relies on wizards and wild cards to compete in the marketplace.

The present discussion of the techno-management imagination recognizes the validity of the heroic-mythic tradition, and accepts its applicability in business, within limits.

By citing the writers of socially constructed reality, I am not endorsing the nihilist element of the postmodern movement. Rather, I am applauding that literature's integration of sociological and psychological aspects of the phenomenon and hoping that more such integration will appear. Advocacy of TMI is not tantamount to encouraging "spirituality in the workplace," although spiritual traditions may be among the keys that allow individual managers to find alternative paths to knowledge.

Key Takeaways

- Techno-management imagination (TMI) has been practiced since long ago as people audaciously tried new methods to deal with unprecedented problems.
- Our mythical imagination affects our choices of technologies, and subsequently, technologies realize and broaden our myths.
- As corporations apply TMI today, managers should balance embracing the unknown against the risk of anti-intellectualism, by generating interdisciplinary skills, better intra-organizational communication, and improved technology absorption capacity.

References

Bradbury R (2004, November 19) To Mars and back. Wall Street Journal Europe, A10

Castañeda C (1970) The teachings of Don Juan: a Yaqui way of knowledge. Simon and Schuster, New York

Churchill J (2004) What's it all about, Lord Russell? The Key Reporter 69(4):2–14

Davis E (1998) Techgnosis: myth, magic and mysticism in the age of information. Three Rivers Press, New York

Hall P (1998) Cities and civilization. Pantheon, New York

Huxley A (1963) The doors of perception. Harper & Row, New York

Kirn W (1991, July). Valley of the nerds. GQ: 96ff

Learner DB, Phillips FY (1993) Method and progress in management science. Socio Econ Plan Sci 27(1):9–24

Muehlhausen R (2004) Corporate innovation: one part genius, nine parts management. The Siemens Round Table on Innovation, Part IV. Wall Street Journal Europe, M8

Nevens MT, Summe GL, Uttal B (1990, May–June) Commercializing technology: what the best companies do. Harvard Business Review, pp 154–163

Phillips F, Kim N (1996) Implications of chaos research for new product forecasting. Technol Forecast Soc Chang 54(1):239–261

Rudolph SE, Lee WD (1990) Lessons from the field. R&D Magazine October, p 119

Salzman M (1991) Laughing sutra. Knopf, New York

Searle JR (1997) The construction of social reality. Free Press, New York

Yeh R, Yeh S (2004) The art of business: In the footsteps of giants. Zero Time Publishing, Olathe, Colorado

Expertise and the Future

10

> *Trust experts, yes, but never about the future.*
> —Matt Ridley, in The Rational Optimist

Sometimes "experts" are good at predicting the future. Sometimes they're not. They may perform more poorly than forecasts crowdsourced from laypersons. Socio-technical changes alter the kinds of knowledge workers, corporations, and governments are seeking to hire—and have irrevocably shifted the center of power in organizations. This chapter looks at these aspects of expertise and the future.

10.1 The Leapfrog Model

Figure 10.1 depicts the four components of scientific progress, in the guise of leaping frogs. As the leapfrog game progresses and Theory is temporarily in advance of Data, scientists enjoy Theory's guidance on where to look for new data. Then Methodology jumps to the front of the queue. And so on. The leapfrog game is the central characteristic of scientific progress.[1] It is how science advances in an ongoing but somewhat unbalanced manner.

Like the microscope and telescope before them, cheap computer memory and sensors (methodologies) have enabled the observation, collection, and recording of vast amounts of new, unexpected data. The methodology frog jumped to the fore, setting the stage for the data frog to overtake it. The problems frog is poised to leap, as industries struggle to hire enough "data scientists" and thus to know their customers better.

One lesson of Fig. 10.1, however, is that there is no "data science." There is only science, and the data frog temporarily leads, even as methodologies for handling large data sets struggle to catch up. Another point is that each time a frog jumps, it

---

[1]See Phillips (2008) for historical examples of each leap.

© Springer Nature Switzerland AG 2019
F. Phillips, *What About the Future?*, Science, Technology and Innovation Studies,
https://doi.org/10.1007/978-3-030-26165-8_10

Methodology

Theory   Data   Problems

Fig. 10.1 The leapfrog theory of scientific advance. Source: Learner and Phillips (1993)

creates a new regime in which different kinds of expertise are needed. In 2019, the need is for data scientists. In 2021, it may be for theorists, or instrumentation experts. Time will tell.

10.2   Knowledge Management and the Knowledge Society

Seismic shifts in our society have begun and are still under way. We are moving from an economy of buying things to borrowing, sharing, and renting. From buying goods to buying services and experiences. From suburban and exurban homes to urban living. (Chapter 11 will tell us we have already passed the point where half the people on the planet live in large cities.) From a Gross National Product based on movement of commodities to a GNP of information flow. From products to "platforms." From analog to digital. From prices rising with inflation to prices dropping due to Moore's Law and global free trade. From army-to-army battlefield or guerilla warfare to cyber-warfare and deliberate terrorist targeting of civilian non-combatants.

Shifts from common broadcast information to customized, narrowcast information, and from bundled to unbundled TV programs, software, and information services, are still in contention, the outcome hard to predict.

To deal with the future, each person must think through the implications of these shifts—what they mean for livelihoods, safety, politics, organizations, and all our other social institutions.

One shift is far bigger, in its propensity to disrupt the economic order, than any that I've just mentioned. It is the shift from a capital-and-muscle economy to a knowledge economy.

In the C&M economy, usually just called the capitalist economy, those with capital called the shots. They alone were able to invest in productive equipment (that's why accountants call it "capital equipment") and hire labor to operate it. Capitalists hired laborers on the basis of the laborers' muscles and ability to do simple but strenuous, repetitive, and often dangerous tasks. As there has almost always been an oversupply of this kind of laborer, workers were easily replaceable. It is different now.

Today—as Peter Drucker (1994), and even, long ago, Karl Marx predicted—knowledge workers hold the economy's reins as much or more so than capitalist investors. The high-tech and healthcare industries, so central to the well-being of the rich countries, depend on highly educated employees. Though we all, mostly, have the same muscles, each of us has distinctive attitudes, skills, and knowledge. For this reason, Silicon Valley executives will tell you that hiring the right person is the toughest obstacle to their companies' prospects. Labor is no longer easily replaceable.

We should not have been surprised, then, when employees of Google—and of several other high-tech firms—insisted that their CEOs should not engage with the newly elected President of the USA, Donald Trump. The CEOs complied. Can you imagine, first, that the power in knowledge-intensive firms now flows from bottom up and not from top down? And second, that a CEO must explain to the board of directors, on which the company's capital investors sit, why he or she cannot control the employees? Indeed, that *control* is an outmoded concept?

As Drucker foresaw, the modern manager's job resembles that of a military commander less than that of an orchestra conductor, just trying to align trusted and skilled people who are doing their own thing. Managers might sigh that they're "herding cats." That should not be a complaint; it's just a description of the nature of today's organizations.[2]

Much has been written about the shift to a knowledge economy, so I'll devote the rest of this section to the forces arrayed against the shift.

Naturally enough, capital investors are reluctant to give up their long-held top spot. Logically enough, they have the cash to mount a defense. They are aided by a long tradition of what Hofstadter (1963) called "anti-intellectualism in America." We can see the 2016 American elections as a conflict between the growing importance of knowledge, versus America's anti-intellectual tradition. The backlash against the knowledge society took form in the election of Donald Trump, and the installation of the anti-globalist Steve Bannon as Trump's chief strategist (Howe 2017).

Silicon Valley needs more knowledge workers, and begs the White House and the State Department for more H1-B visas for educated immigrants. Curiously and probably tragically for America—and I leave it to conspiracy theorists to draw any possible connections—higher education in America gets ever more expensive, its graduates crushed under student loan repayments. At the same time, European countries make higher education free or nearly so.

---

[2]As an educated but not wealthy writer of books, I greet the new importance of people like myself with some glee. However, this section is not a value judgment of capitalism. Capitalism has brought us much benefit, though frequently with notable cruelty. I am not here to argue whether the benefits have exceeded the harm, over the course of history. I am here to say, along with Drucker, that capitalism will not carry us through the century to come, and will not solve the global ecological crisis.

The USA still has fine universities, but K-12 education is distressed. Not just countries like Finland and Taiwan, but others like Costa Rica—which until recent decades was truly a "banana republic"—have K-12 outcomes superior to America's. US universities admitting students with poor primary/secondary preparation must divert resource into remedial courses. At the postgraduate level, US universities depend on well-prepared foreign students for the most innovative work.

The People's Republic of China, having a strategic interest in all things American, continues to send good students to the USA. Parents in other, particularly Asian, countries fear their offspring will fall prey to drugs, crime, and racial bias in America, and send their kids to Australia, Canada, and the UK for school. In any case, in the current clime, these youngsters would have a hard time getting a US student visa. All this does not bode well for the future of innovation in the USA.

10.3   Limits of Expertise

Income inequality and wealth inequality in the USA continue to grow. The rich, now very rich, have ample resources to defend the capital-dominant model against the knowledge-dominant model. The backlash against the knowledge economy will last a depressingly long time, though the ultimate outcome is in little doubt. Knowledge-able people will have to grin and bear the epithets "egghead," "snowflake," and so on, for some time to come. Expertise has become devalued, unfairly and harmfully for the public good.

Yet expertise does have sharp limitations. Wu and Dunning (2018)[3] write of *hypocognition*, a state of not being able to comprehend something one has never before been exposed to:

> facing a concept we cannot fathom, an emotion we cannot grasp, a idea that arouses in others fervor and enthusiasm but strikes us as bizarre, a principle that must, against our own reason, be unreasonable. Who are most likely to fall prey to hypocognition? Experts who are confined by their own expertise. Experts who overuse the concepts salient in their own profession while neglecting a broader array of equally valid concepts.

Who is an expert? There has always been a problem of definition. When two experts disagree, does it mean one of them is *not* an expert? Is the louder pundit the greater expert? Does an advanced degree make you an expert? I suppose it's "no" to all these questions.

Are experts corruptible, for example, fudging lab results when funding comes from corporations affected financially by those results? Are true experts in scientific domains susceptible to dominant personalities and other group dynamics effects when participating in expert panels, thus biasing their statements? Unfortunately, the answer to these questions is "yes." The American Association for the Advancement

---

[3]For more on this, including remarks by Feynman and Kahneman, see http://www.futurecasts.com/ J)%20Tetlock%20&%20Gardner,%20Superforecasting%20II.htm

of Science (AAAS) and other professional bodies are urgently devising fixes for both these problems. The fixes won't be soon in coming, and may not work.

How to choose experts who can help us push our vision farther into the future? This is still very much more an art than a science.[4]

Thus, we are reduced to defining "expert" as *someone who you think knows more than you do.* (Or, to be fair, someone who can creatively put together pieces of knowledge better than you can.) On this basis, experts are frequently recruited to Delphi panels, to refine future scenarios, and to render opinions like those described in the next section.

10.4   An Expert View of 2052[5]

In 2017, the 350-year-old German firm Merck KGaA surveyed readers of the journals *Science, Nature,* and *Harvard Business Review,* asking their views of how the business and technology environments would evolve in the next 35 years. The firm and the journal publishers considered their readers to be experts—their expertise maybe having arisen before reading the magazines, or maybe because of reading them!

Questions addressed the readers' expectations about the future, and their desires about the future. *Science* attracts more American readers and *Nature* more readers in Europe. Both appeal to scientists and engineers, while HBR's readers are predominantly business people.

These groupings offered my company, which analyzed the survey data, ample opportunity to contrast the groups: scientist vs. businessperson, America vs. Europe vs. elsewhere in the world, senior vs. junior career stage, and expectations vs. hopes. My company also asked the same questions of the editorial board of the journal *Technological Forecasting & Social Change,* who are experts in technology forecasting.

An early and striking lesson was that *between-group differences in views are substantial and really more interesting than the areas of agreement.* Another was that, despite the group differences, each group is very large, and most respondents agreed that *large groups of people articulating a similar anticipated future are significantly influential in making that future happen.* The implication is that even without a central operational plan in hand today, constructive steps will grow organically, as a result of the group vision.

All groups expect medical and computer sciences to show the most potential for breakthrough advances in the coming 35 years. All groups also saw much potential in materials science, knowing that much of the technology we rely on depends in turn on new, advanced materials.

---

[4]See Stefanie Mauksch et al. (2019).

[5]This section is based on Betz et al. (2019).

On the technology front, and again in agreement with the TF&SC board, business people expected the most dramatic advance in artificial intelligence, closely followed by medicine/healthcare, advanced analytics for big data, robotics, energy technologies, and synthetic biology and genetic engineering. Business people differed from the science group by having much higher expectations for space travel.

In the spirit of bounded futures (see Chap. 5), we asked the TF&SC board what advances they do NOT expect to see in 35 years. The answers: commercial space stations, a secure Internet, and reversing the effects of climate change.

As regards needs and desires, business people most often named computerization, healthcare, transportation, and environment/food. They added many imaginative and sometimes whimsical items in this category. These included vaccines for autoimmune diseases, hydrogen-fueled supersonic transport, personal flying, detection of civilizations on other planets, and sex robots.

All expect new cross-pollinations across scientific fields, specifically computer sciences with humanities and social sciences, and biological and medical sciences with computer sciences. In the same way, they expect technological convergence of big data and robotics with AI, and synthetic biology/genetic engineering with medicine and health. This means the boundaries between traditionally defined industries will continue to disintegrate.

Another difference to be considered is that global R&D funding trends show the greatest increases[6] in aerospace, automotive, and information/communication tech. These categories do not match our respondents' forecasts and priorities. We have to remember, though, that R&D funding changes with each government administration and for this reason cannot be forecasted more than 2 or 3 years ahead.

Notably, no respondents mentioned weaponry. Doubtless many new and terrible weapons will be developed and deployed in the coming 35 years. It seems readers of these journals simply do not think about weapons in the course of their daily lives, or wish to.

Business people may be more susceptible to journalistic and advertising hype (see Chap. 8) than their counterparts in science and forecasting, and more bedazzled by the many impressive gadgets that have come to market in the last decade. This would explain why the scientists and forecasters were more cautious than the HBR readers in predicting bold advances in the sciences.

It seems that a responsible futurology would explore, through dialog or psychological testing, *why* the various groups' enthusiasms differ as they do. Only then can we look at whether the various views can or should be reconciled, and what to do in either case.

---

[6]Global R&D funding trends, 2014–16, https://www.iriweb.org/sites/default/files/2016GlobalR%26DFundingForecast_2.pdf

Key Takeaways

- Among many seismic shifts happening in our society, the shift from a "capital and muscle" economy to a knowledge economy is likely to signify a reordering of power in organizations. However, reactionary forces are slowing down this shift in America.
- One large expert survey shows medical and computer sciences (underpinned by advances in new materials) are expected to produce the most potential for scientific breakthroughs in the next 35 years. The most potential for technological advance is expected to lie in the areas of artificial intelligence and medicine/ healthcare, with much cross-pollination between the two.
- Experts, who are highly susceptible to hypocognition and bias, and easily as corruptible as anyone else, are sometimes no better than laypersons in predicting the future.

References

Betz UAK, Betz F, Kim R, Monks B, Phillips F (2019) Surveying the future of science, technology and business—a 35-year perspective. Technol Forecast Soc Chang 144:137–147

Drucker P (1994) Post-capitalist society. Harper Business, Reprint (Kindle) edition

Hofstadter R (1963) Anti-intellectualism in American Life (3rd printing). Knopf

Howe N (2017, Feb 24) Where did Steve Bannon get his worldview? Washington Post. Retrieved from https://www.washingtonpost.com/entertainment/books/where-did-steve-bannon-get-his-worldview-from-my-book/2017/02/24/16937f38-f84a-11e6-9845-576c69081518_story.html?utm_term=.8c72c47d9856&wpisrc=nl_most-draw8&wpmm=1

Learner DB, Phillips FY (1993) Method and progress in management science. Socio Econ Plan Sci 27(1):9–24

Mauksch S, von der Gracht H, Gordon TJ (2019) The need and search for expertise in foresight. Technol Forecast Soc Chang, 50th Anniversary Special Issue

Phillips F (2008) Change in socio-technical systems: researching the multis, the Biggers, and the more connecteds. Technol Forecast Soc Chang 75(5):721–734

Wu K, Dunning D (2018) Unknown unknowns: the problem of hypocognition. Scientific American, https://blogs.scientificamerican.com/observations/unknown-unknowns-the-problem-of-hypocognition/

The Present and the Future

11

*Twenty years from now you will be more disappointed by
the things you didn't do than by the ones you did.*
— Mark Twain

The Economist reported that in early 2007, for the first time in history, more humans lived in cities than in the countryside. We are now a different species, in terms of the environmental niche we inhabit. One thinks of Isaac Asimov's Trantor, the planet that was completely covered by buildings. Is Earth headed for a similar future?

2004 was the first year Amazon.com moved more dollar volume in consumer electronics than in books. Investors had to revise their image of the company and ask new questions of its management. When Toyota overtook GM in car sales in 2007, it raised questions of US competitiveness and Japanese market dominance that had not been asked since the early 1980s.

Crossovers of this kind are dramatic. They are not surprising, because they happen in the middle of the well-known S-shaped life cycle curve (see Fig. 11.1), where the growth or decline of a behavior has become steady and predictable. Any astute analyst can see them coming, well in advance. Crossovers are no less thought-provoking, though, for all their individual predictability.

And when many significant crossovers (in varied social and market domains) occur in fast succession, they say something important about our times, and about market opportunities.

These crossovers often signify important trends. When it emerged that more US households now keep pet cats than pet dogs, it suggested that more households were becoming apartment dwellers, with all that implies for appliance sales and space-saving storage devices (not to mention pet care products). When a trend captures 50% of its domain, it's a sign that it is here to stay (Phillips 2007). For this reason, crossovers tell us important things about our future and usually show us that the future is already here.

© Springer Nature Switzerland AG 2019
F. Phillips, *What About the Future?*, Science, Technology and Innovation Studies,
https://doi.org/10.1007/978-3-030-26165-8_11

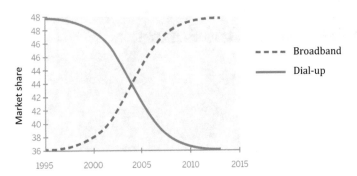

Fig. 11.1 Fitted trajectories of two modes of Internet access, illustrating their crossover. Source: Author

Each crossover involves a new behavior overtaking an old behavior. This might be because the incidence of the old behavior is declining (as in music sales through traditional channels), or because both are growing but the new behavior is growing faster than the old one. The year 1991 provided an example of the latter: investments in both IT and traditional capital goods were growing, but IT investment was the star performer by far.

Many of the crossovers listed below illustrate a new behavior *substituting* for an older one. (That is, an individual engaging in more of the new behavior necessarily engages in less of the old one, and both behaviors deliver the same benefit.) Dial-up vs. broadband home Internet connections are a classic instance of technology substitution. Let's have some fun with that one before returning to the (really more interesting) question of why the others are *not* classic cases of substitution.

The crossover for home Internet connections happened in 2004. With just a little extra data, we can fit the Fisher-Pry equation[1] for technology substitution. From CNET (http://news.com.com/2100-1034_3-5815756.html) we learn that "29 percent of North American households connected to the Net via broadband." Home broadband penetration in Canada was then higher than in the USA, but let's go ahead and apply the 29% figure to the USA. Next, websiteoptimization.com/bw/0607/ tells us that between March 2005 and March 2006, the percentage of broadband households in the USA grew from 34.97 to 44.45%.

Knowing that broadband and dial-up had equal market shares in 2004 (the crossover point), the total market for both types of Internet access was twice 29%, or 58% of US households. In 2005, 34.97% penetration meant broadband had 34.97/0.58 = 60% share of net access, and in 2006 broadband's share was 44.45/0.58, or 77%.

The Fisher-Pry equation is $\log(s_{n-1}/s_n) = kt$, where s_{n-1} is the market share of the old technology, and s_n is the market share of the new technology. k is a parameter, and t is time. Fisher-Pry is a single parameter model. The bad news is that makes it

---

[1]Fisher and Pry (1971).

rather rigid, and inescapably symmetric. The good news is that our three data points are more than sufficient for fitting this simple model. Skipping some details, we arrive at the fit shown in Fig. 11.1.

According to CNET, Forrester Research predicts 62% broadband penetration by 2010, the newcomers attracted mainly by lower prices. How does our Fisher-Pry forecast square with Forrester's prediction? Our curve shows a 96.6% share for broadband in 2010. Multiply by the full market size (at least this was the full market size in 2004, though it will have grown by now) of 58% of US households, and we arrive at a competing estimate of 56% penetration for 2010.

This is a bit less than Forrester's forecast, and the difference seems partly due to our inclusion of the Canadian data. Forrester also says declining prices will drive higher broadband penetration. The Pew Trusts disagree. Their 2004 survey[2] of reasons for switching to broadband at home showed 36% of switchers were moved by "connection too slow or frustrating." Only 3% mentioned price as a motivator. As you form your own opinion (I said this was for fun), let's look at some of the other crossovers.

No one would suggest that IT is a substitute for capital equipment. They don't deliver the same benefit; we still need manufactured goods that conventional computers cannot make for us. Some US companies invest in more IT domestically as they offshore manufacturing to China. This example illustrates the complexity that can make the simple curve of Fig. 11.1 an inappropriate tool for analyzing crossovers.

When Japan's pop culture exports passed its car exports, it said less about Japan's industrial structure than about new competition for Hollywood and Bollywood. It would be a mistake to view this as an instance of substitution.

It is usually said that a technology is "mature" when the rate of new patents in the area begins to level off. When pornography loses its place as the killer app for a new communications technology—in 2007 social networking sites overtook porn to become the killer app of the World Wide Web—that might be a truer measure of the technology's maturity. And, I guess, the customers' maturity too.

In 2001, California's population of children aged 5 years and under numbered just over three million. Forty-eight percent of these children were Hispanic.[3] The crossover has surely occurred by now, and it has immense implications for the future composition of the workforce, for educational programs, for protection of children (who are themselves US citizens) of illegal immigrants, and for many other socio-economic issues. Small wonder that writer-director Wayne Kramer's 2007 movie *Crossing Over* is a "multi-character canvas about immigrants of different nationalities struggling to achieve legal status in Los Angeles" (imdb.com).

I first became interested in crossovers when Market Research Corporation of America, for whom I then worked, publicized the new ascendancy of cats over dogs

---

[2]Pew Internet and American Life Project, February, 2004, www.pewtrusts.com/pdf/pew_internet_broadband_0404.pdf

[3]http://www.ppic.org/content/pubs/jtf/JTF_YoungChildrenJTF.pdf

as household pets. The event was intriguing not only for its own sake, but because the pet ownership checkbox on our demographic questionnaires was so tangential to MRCA's core business of product purchase tracking. An alert (or bored) analyst had decided to tabulate these checkboxes, and accidentally discovered the crossover. An equally alert manager issued a press release, which the media immediately picked up. Crossovers are newsworthy!

Keep your eyes open for crossovers, and understand the problems and opportunities they represent. I keep my collection posted at www. generalinformatics.com/crossovers.html. If you would like to suggest additions, send them to me at info@generalinformatics.com.

Incidentally, in 2008 social networking overtook e-mail in "number of Americans using," according to Neilsen. Of course "search" remains the king app of the net, but social networking is now not far behind.

Here is a collection of crossovers, past, present, and forecasted (with their sources noted). Peruse them. Imagine—and, if you will, exploit—the trends they imply.

11.1 Urbanization and Households

1984 MRCA Information Service's consumer purchase tracking service first published the fact that, for the first time, **more U.S. households owned cats than dogs**. This was a sign of greater urbanism and trends toward apartment and condominium living. It was a sign to marketers that the future lay more in miniature and under-the-cabinet kitchen appliances, and less in tractor mowers.

2000 **more *salsa* was sold** in the U.S. than ketchup.

2004 the number of U.S. households with **broadband access equals those using dialup** Internet access.

2006 Debit cards grew steadily, hitting 50% in 2006. Today, there are **more debit cards than credit cards** in circulation. *Austin American-Statesman*, October 02, 2011.

2007 According to *Wired*, the **majority of human beings live in cities**.

2013 "The average American will spend **more time with digital media** each day (about 5 hours) **than watching television**." (eMarketer)

2013 **Adult diapers outsell babies' diapers in Japan**. (The Lowdown. Posted: Jul 13, 2013; 09:02 AM PDT)

2014 April, According to a Nielsen study, half of Americans over 55 years of age now own a smartphone. This is a 10% increase from just earlier in the year, making this the **first time the majority of all age groups have owned a smartphone**.

2014 At the end of June 2014, the number of **people subscribing to broadband internet** from the nine largest US cable companies (49,915,000) **exceeded the number of television subscribers** (49,910,000) for the first time. That's according to a tally by Leichtman Research Group.

2014 **More women than men own driving licenses** for the first time in automotive history. The LowDown.

2015 America's **middle class shrunk to less than 50% of the population** for the first time since the early 1970s. *Financial Times.*

2015 We now **spend half of our waking lives consuming media**. *Quartz*

2015 "Americans are **buying more pot than Girl Scout Cookies**. In 2015, medical and recreational marijuana posted between $3 billion and $3.4 billion in revenue, while the more wholesome confections collected $776 million in cookie sales." Qz.com.

2017 Suburban poverty is higher than ever before. For the first time, the **number of poor Americans living in suburban areas has surpassed the number of poor city dwellers**. (Brookings Institution)

2020 Before 2020, **chicken will overtake pork** as the most-consumed meat on the planet. (*The Economist* 9/14/2013)

11.2 Culture and Society

1990 Americans first made **more visits to "alternative health providers" than to conventional M.D.s**. The next half-decade saw explosive growth in herbs, vitamins, and natural food stores.

2004 California's **spending on prisons passed spending on universities**. *The Economist* 9/10/11, p. 36.

2004 the **opening weekend gross of a new video game** exceeded the expected opening weekend **gross of a new major movie.**

2006 **Exports of Japanese pop culture** items—video games, comic books, etc.—**exceeded Japanese automobile exports**, according to National Public Radio. So much for American dominance of worldwide pop culture. Watch out Hollywood!

2007 Chat, discussion, and **social networking web sites enjoyed more hits than pornography sites**, according to *The Economist.*

2007 According to Bill Gates, speaking at COMDEX, "Young people" are spending **more hours in front of their PCs than in front of their television sets.**

2007 According to Pew Research, the values and **purchasing behavior of middle- and upper middle-class African-Americans** was more similar to those of white Americans than to those of low-income Blacks.

2008 the **U.S. overtook France as the world's leading wine-drinking nation**. Growth in U.S. consumption is driven by Australian wines (*Business Week* prediction, Feb 28, 2005, p. 14)

2008 Brookings Institution: "For decades, the **entrance of firms** outpaced their **exit**, meaning a net increase in new businesses. The authors see that. . . as a proxy to an inclination toward entrepreneurism. But since at least 1978, the lines have converged, albeit slowly. In 2008, they reached a watershed moment and crossed."

2009 **more people worldwide used the Internet for social networking than for email**. Of course, the primary use of Internet continued to be "search."

2009 "**Drugs exceed motor vehicle accidents as a cause of death**." *Austin American-Statesman*, 9/18/11.

2010 The news that **student debt has surpassed credit card debt** as the largest source of consumer debt in the United States is a function of rising costs of attending higher education, cuts to state and federal financial aid, and the growth of for-profit private industry around the student loan bubble.

2010 **Student loan debt** eclipsed the total amount that Americans owed on their **credit cards**. (*New York Times*)

2011 In US, **more births of non-Caucasian babies than Caucasian** babies (*The Economist* May 19th, 2012, p. 8)

2011 "For the first time in recorded history, the Latin America and Caribbean region was home to **more people in the middle class than in poverty**. " *Site Selection International*, December 2014.

2011 **China** has overtaken the **U.S.** to become the world's **biggest producer of PhDs** (*Nature* 472, 276)

2012 More British soldiers and veterans **took their own lives than were killed in battle**. The 50 suicides exceeds the 40 soldiers who died fighting the Taliban in Afghanistan during the same period. http://www.independent.co.uk/

2013 **More people have cellphones than toilets**. UN reports there are now more people with mobile phones (six billion for world population of seven billion) on earth than there are with access to clean toilets (4.5 billion) (Matthew Yglesias Friday, March 22, at 2:21 PM).

2013 For the first time, **China**'s share of global CO_2 **emissions** surpasses **European Union's**. *Science* reported this announcement from the Global Carbon Project.

2014 The amount of **money invested in defined-contribution pension plans will exceed the amount invested in defined-benefit** schemes. (The number of people enrolled in defined-contribution schemes surpassed the number in traditional defined-benefit plans in 1988.) *The Economist*, June 12, 2008. http://www.econo mist.com/finance/displaystory.cfm?story_id=11529345

2014 For the first time, the **craft beer** industry shipped **more barrels** of beer **than Budweiser** last year. Matt Schiavenza reports in *The Atlantic* November 28, 2014.

2014 "Stephen Marche wants the U.S. to wake up to a serious public health issue. 'Two years ago, **suicide became the leading cause of death by injury in America, surpassing car accidents** for the first time.' And the major reason for that change was a cohort shift: Men and women between the ages of 35 and 64 are increasingly committing suicide. The latest addition to these statistics is Robin Williams. Since nothing ever happens in America until it happens to a celebrity, perhaps this will be the moment when we notice that we're living in the middle of a suicide crisis." *.micCheck*

2014 The *Wall Street Journal* reported a record: Almost **half of all Americans** (47%) now say they **want the United States to be less active in world affairs.** A decade ago, only 14% felt this way.

2014 For the first time, the U.S. stopped **more non-Mexicans than Mexicans** at the border (*PolicyMic*)

2014 U.S. law enforcement **officials took more property** from people **than burglars did** (*MicCheck Daily*)

2014 The percentage of **women with a bachelor's degree in the US (32%) surpassed the percentage of men (31.9%) with one,** according to data from the US Census Bureau.

2014 Americans now have nearly **as many smartphones as TVs** (Recode.net), and Bloomberg adds that as of **2014**, we spend more time staring at phones than at TVs.

2014 **More Harvard Business School grads went into technology than into banking** for the first time since the dot-com era. (Quartz).

2015 College students who **smoke weed now outnumber those who smoke cigarettes.** *MicCheck Daily.*

2015 Corporations now **spend more lobbying congress than taxpayers do funding it.** *The LowDown.*

2015 "The **UK** is already one of the very few countries in the world where **non-academic staff already outnumber academics** .... directors of corporate affairs and human relations and the heads of research excellence framework strategy, overseas recruitment, research impact, fundraising, external relations and brand management." https://www.timeshighereducation.com/content/laurie-taylor-on-academics-v-administrators

2015 **Foreign** born college grads **working in the US now earn more than US college grads,** *Jordan Yadoo reports in Bloomberg.*

2015 Adjuncts or lecturers or other species of **contingent labor now account for more than half of all post-secondary teaching positions** in the country. http://www.psmag.com/business-economics/pay-for-decent-teachers-not-dr-phil

2015 **Employment in intangible industries has exceeded employment in tangible industries** (Ken Jarboe reports in The Intangible Economy blog). "Tangible activities are primarily physical; intangible are primarily mental. Cutting hair, ringing up a sale at a cash register, making a car, harvesting a crop—all of these are primarily a physical activity. The transaction involves the movement of atoms. Designing a poster, negotiating a deal, writing an article for the *Washington Post*—these are primarily mental involving the manipulation of intangible information bits."

2015 **There are now as many solar jobs as coal jobs** in the US. *Vox Jan 18.*

2015 **Millennials overtook Boomers** as the biggest share of the US workforce.

2015 For the first time in human history, **media rights for sporting events are about to surpass gate receipts**—e.g., the box office—in financial value. *The LowDown Report.*

2015 For the first time in history, **more people die today from eating too much than from eating too little**; more people die from old age than from infectious diseases; and more people commit suicide than are killed by soldiers, terrorists and criminals combined (Harari 2015). Yuval Noah Harari (2015) *Homo Deus.* Vintage London.

2016 There are now **more poor people in America than in China**. (MicCheck/ Credit Suisse Wealth Report).

2016 According to Senator Elizabeth Warren, **support for Donald Trump from KKK members exceeded that from Republican Party officials** in early May.

2016 **Guns and cars now kill Americans at the same rate**. Motor vehicle-related deaths have dropped by about 25% in the past decade, while gun deaths have mostly remained flat. (MicCheck)

2016 **Service sector surpassed manufacturing** in **China**. (Chinese Academy of Sciences)

2016 We're now closer to **gender parity in education** than at any time in modern history. Countries including Botswana, Nicaragua and Costa Rica have fully closed their enrollment gender gaps, and Nepal, Rwanda and Zimbabwe are 90% of the way there. (Devex)

2016 The rate of new business formation in the USA has fallen by 50% since 1978. Today, there are **more deaths of businesses than there are new business** births, according to Jeffrey Pfeffer, Professor of Organizational Behavior at the Stanford Graduate School of Business. (*Forbes*)

2016 Tables Turned: **Vinyl records outsell digital downloads** for first time in the UK. (The Guardian)

2017 Nearly Half of Millennials and Gen Xers **don't watch any traditional TV**. (*AdAge*)

2017 **China surpasses the US as the world's leader in scientific output**, due to increased Chinese biomedical publications and reduced US support for research in that growing field. (Mic.com) https://mic.com/articles/180075/good-going-america-china-is-about-to-surpass-the-us-as-the-worlds-leader-in-science#. mtgnpoBpz

2017 **Trump's approval rating drops below his disapproval rating**, according to a Reuters/Ipsos tracking poll. (IPSOS)

2017 In America, **more people now die from antibiotic-resistant bacteria (superbugs) than from homicides**. (*Foreign Policy* magazine)

2018 China to become the world's largest economy. Thanks to four decades of strong economic growth, **China is expected to overtake the U.S.** as the **world's biggest economy** in 2016. Researchers forecast rapid growth in China in the coming years, which will gradually slow as the country becomes more industrialized.

2025 "There are **more obese people than underweight people** in the world. According to a new study published in *The Lancet*, Earth's obese population has increased sixfold since 1975, and if current trends continue, one in five adults worldwide will be obese by 2025."

2030 The United Nations reports that by 2030 **India will have more people than China**. (World Economic Forum). https://www.weforum.org/agenda/2016/06/when-will-india-have-more-people-than-china/

2043 More than **half of USA residents will be** "**nonwhite**," *Austin American Statesman,* April 6, 2014.

2050 "By 2050, **plastic will be more plentiful in our oceans than fish**. As humans continue to increase the production of plastic while remaining terrible at recycling it, the material threatens to outweigh marine life pound-for-pound." *USA Today.*

11.3   Technology, Investment, and Trade

1991 U.S. **investment in** the tools of the information trade—**computers and communications gear—outpaced capital spending** in the industrial sector. By 1992, only one year after the two trend lines had crossed for the first time, capital investment in information technology was nearly $25 billion higher than traditional industrial capital investment, and pulling ahead quickly. (David Kline, "Market Forces," *HotWired* 12/11/95)

1995 **More PCs were sold than television sets**. This implied at the time that the "set-top box" for home delivery of interactive multimedia would be a dead-end technology. However, the post-2004 growth of HDTV sets uncrossed this crossover.

2001 **More women than men shopped online**. Women made up 58% of the 29 million online shoppers in the Thanksgiving-Christmas weeks of the year, according to the Pew Trust, making the mix of online shoppers nearly identical to the mix of bricks and mortar shoppers. As most early adopters of anything—including online shopping—are young males, this crossover indicated a new maturity of the World Wide Web.

2002 **Mobile phone subscriptions outstrip land lines** (*The Economist*)

2003 **More digital cameras were sold than film cameras.** This was about 5 years sooner than forecasters expected. Thus Kodak was described in the press as "the troubled imaging company."

2003 NPD Group reported that **more laptop computers were sold** in stores **than desktop computers**. (Obviously this excludes online and b2b sales.) NPD also told us that in the same month, for the first time, Liquid Crystal Display monitors accounted for a majority of monitor sales in stores, displacing cathode ray tubes.

2004 Sales of servers based on **Intel and AMD chips exceeded sales of UNIX servers** (*Business Week*, 2/14/05, p. 81)

2004 **Amazon.com's home electronic sales volume exceeded its book sales** (December)

2005 **Computer models execute more than half of all stock trades**. *Business Week* 4/18/05, p. 88.

2006 **Video game sales** exceeded **music sales** (PricewaterhouseCoopers).

2007 **Toyota sold more cars than General Motors**. This makes Toyota the world's top auto maker by unit sales, a spot GM had held since 1931.

2007 **Japan's exports to China overtake those to US** (*The Economist*)

2009 **VCs invested more in healthcare startups than in IT startups.** *Wall Street Journal*.

2010 **China** files **more patent applications than Japan** (*The Economist*)

2010 **China overtook the US in manufactured output, energy use, and car sales**. (The Economist 4/21/12, p. 41)

2011 Cell **phones outnumber humans** in the US, according to the wireless trade group CTIA.

2011 Sales of **digital music surpass sales of all physical music** (CDs, vinyl, etc.), according to the research firm Strategy Analytics.

2012 "Readers are voting with their wallets: The eBook is winning. In the US, **eBook sales are now topping hardcovers** for the first time (story in TechCrunch). Not everywhere of course. According to the Bowker Global eBook Research, the global market for eBooks is driven—in that order—by India, Australia, the UK and the United States. The laggards are Japan and (no surprise) France." *Off The eBook Shelf*, June 18, 2012,10:04 am | Edited by Frédéric Filloux

2012 Financial Times' **Digital Subscribers Surpass Its Print Readers**. The Financial Times' digital subscribers surpassed its print customers for the first time, marking a milestone that the *New York Times* and other newspapers with so-called paywalls are aiming to reach. businessweek.com

2012 It happened for the first time: **Bot traffic eclipsed human Internet traffic**, according to the bot-trackers at Incapsula.

2012 **Bikes officially outsell cars in European Union**. An analysis by National Public Radio indicates that in 2012, more bikes than cars were sold in each of the 27 members of the European Union except Belgium and Luxembourg.

2013 Sales of **smartphones exceed those of non-smart** ones. (IHS Research)

2013 marked a turning point for solar photovoltaics, according to Clean Edge's Clean Energy Trends 2014 report, as newly installed **solar PV generating capacity exceeded that of wind power** for the first time. The solar PV market grew to $91.3 billion from $79.7 billion in 2012, with a record 36.5 GW installed globally.

2013 Bloomberg: **China** spent more on **smart grids** than the U.S. for the first time in 2013, with the $4.3 billion it invested accounting for almost a third of the world's total.

2013 According to ComScore and Morgan Stanley Research**, the number of mobile internet users surpassed desktop internet users.**

2014 Credit card spending by non-Koreans visiting Korea outstrips card spending by Koreans on overseas trips. *Korea Herald,* Nov 29–30, 2014.

2014 January **2014 Mobile usage has overtaken desktop usage.** In addition, Flurry, which tracks app usage, reports that **apps account for the vast majority of time spent vs. the mobile web.**

2014 **More mobile traffic than PC traffic.** Marissa Mayer, Chief Executive Officer, Yahoo, USA; Davos Co-Chair in The New Digital Context.

2014 **Texas overtakes California as #1 technology-exporting state**, according to a tech industry lobbying firm.

2014 BookBaby reports: Over **50% of all email gets opened on a phone**.

2014 The **U.S. becomes world's biggest oil producer** after overtaking Saudi Arabia and Russia as extraction of energy from shale rock spurs the nation's economic recovery. Bank of America Corp.

2014 **2014 Solar met more than 50% of Germany's total electricity demand** for the first time this week. *Triple Pundit.*

2014 The amount paid out to **iOS app developers** in 2014—over $10 billion—now exceeds **Hollywood at the box office**. *Business Insider.*

2014 Number of WhatsApp Messages Now Exceeds Standard Texts. *The Economist.*

2015 Teenagers **interact with friends over social networks** just about as much as they see each **other in person,** the Pew survey shows. http://www.psmag.com/nature-and-technology/my-best-friend-is-named-dogpound99

2015 **Uber has overtaken taxis** as a preferred mode of transportation among road warriors in Q2, according to Certify's analysis of second-quarter expense reports. *Nancy Trejos reports in USA Today.* Ubers overtook yellow cabs in terms of the number of vehicles in 2015, and in the number of passengers in July this year. (*Nikkei Asian Review*)

2015 McGraw-Hill: A **majority of** college students' **studying is taking place via mobile device.**

2015 More than half of U.S. POS terminals to be EMV chip-enabled by year-end (Aite Group)

2015 **Natural gas overtook coal** as the top U.S. power source. (*The Lowdown*)

2016 By 2016, **over half the world's Internet traffic** came from **Wi-Fi connections**. Juniper research has forecast that Wi-Fi networks will carry almost 60% of smartphone and tablet data traffic, reaching over 115,000PB (Petabytes).

2016 **Uber and Lyft have overtaken rental cars** among business travelers (Bloomberg, 4/**2016**)

2016 "This [**2016**] is the first year in history where **investments in renewable energy** have outpaced those in **fossil fuels**. So the market is moving ahead much, much faster than most people understand," John Morton, the White House's senior director for energy and climate change, said (*Washington Post*).

2016 China overtakes USA in **supercomputing power** (*Science*)

2016 **Chromebooks outsold Macs** for the first time in the first quarter of this year. (CNN Money, May 20, **2016**)

2016 Japan now has **more electric charging sites than gas stations**. (World Economic Forum)

2016 **Audiobooks** are starting to **outsell print books.** (Berrett-Koehler publishers)

2016 Within 4 years **solar will be cheaper than coal.** (BNEF). https://assets.bwbx.io/images/users/iqjWHBFdfxIU/ivDyNs7jo1rs/v2/800x-1.png

2016 In the mid-1990s, the best voice recognition software suffered error rates of about 95%. This fall, a team of Microsoft engineers published a paper documenting that **robots hear better than you do**. The researchers wrote that "our automated system establishes a new state-of-the-art, and edges past the human benchmark." (*Kansas City Star*).

2017 "Internet of Things." The number of **things** connected to Internet equals the number of connected **people**. (Korea Ministry of Science, ICT, and Future Planning).

2017 **Digital ad spending** surpasses TV spend.

2017 Global **spending on mobile ads surpasses desktop spending** for the first time, according to Zenith, the research arm of ad giant Publicis Groupe. Not only will mobile viewing increase significantly, non-mobile will decline for the first time. (The LowDown)

2017 Analysis: UK auction reveals **offshore wind cheaper than new gas**. (CarbonBrief). https://www.carbonbrief.org/analysis-uk-auction-offshore-wind-cheaper-than-new-gas

2017 Solar power is pacing to make it easier for the world to dump coal once and for all. Within the next 4 years, **solar power** will finally be **cheaper than coal** nearly everywhere, aligning economic and environmental interests. (*Futurism*)

2017 **Amazon** surpasses **Macy's** as the **largest seller of apparel in America**, with big implications for jobs and shopping districts. (*New York Times*)

2017 **Tesla's market cap** exceeded those of **Ford and GM**. (*Business Insider*)

2017 The price of **one Bitcoin** climbed above the price of **one ounce of gold** for the first time ever. (*Business Insider*). http://www.businessinsider.com/bitcoin-price-tops-gold-price-2017-3

2017 374,000 people were **employed in solar energy**, while coal, gas and oil power generation combined had a workforce of 187,000. The boom in the country's solar workforce can be attributed to construction work associated with expanding generation capacity. The gulf in employment is growing with net generation from coal falling 53%. Solar energy added 73,615 new jobs to the U.S. economy over the past year while wind added a further 24,650. In the United States, more people were employed in solar power in 2016 than in generating electricity through coal, gas and oil energy combined. (*Forbes*)

2017 **China** surpasses **U.S.** as world's **biggest movie market**. (*Los Angeles Times*)

2018 **Open access** articles overtake **subscription articles.** (World Economic Forum)

2022 **Wind power** to be **cheaper than gas** power, 2020 or 2022. Source: https://3.bp.blogspot.com/-O3dmSk6Cq2E/WgnGWWU54aI/AAAAAAAAwHY/TxRYBM6nlqcCu3TmbqF6sslKq-wSPkA6wCLcBGAs/s1600/Image%2B-%2Bwind1.jpg

2035 The number of **Muslims in Great Britain** will exceed the **number of Christians**, according to a British religious think-tank. http://www.dailymail.co.uk/news/article-564722/More-practising-Muslims-Christians-Britain-2035.html

Key Takeaways

- Crossovers, in which an old behavior is overtaken by a new behavior, provide us with noteworthy suggestions about the future.
- The sheer number of recent crossovers, in areas of importance to our lives, underscores how quickly our world is changing.
- Besides classic illustrations of substitution, there are diverse cases of crossovers in which two behaviors do not necessarily deliver the same benefit or involve substitution for each other. When perusing crossovers we should try to understand the real problems and opportunities they embody.

References

Fisher JC, Pry RH (1971) A simple substitution model of technological change. Technol Forecast Soc Chang 3:75–88

Phillips F (2007) On S-curves and tipping points. Technol Forecast Soc Chang 74(6):715–730

Policy and the Future

12

> *I think anyone that tries to predict more than five to ten years*
> *ahead is a bit of an idiot.*
> *So many things can change unexpectedly.*
>
> —James Lovelock

In this chapter, we'll look at three public policy issues that most intimately relate to the future. (There are more, but these are the ones that move my pen at the moment.) Future-oriented technology assessment attempts to predict the impact of a new technology on society and on markets. An assessment of aspects of autonomous vehicles illustrates the complications that tech assessment addresses.

A second public policy issue is the question of free markets (a notion that in today's US political climate implies an opposition to government planning). The third is sustainability, obviously of great import for the future. Finally, the chapter mentions the awful things, like wars, that can disrupt and ruin the best plans and policies.

12.1 Technology Assessment

A chain saw, sporting all the safety interlocks, might still kill you if you use it carelessly. You're self-confident and you suffer the usual optimism bias. You buy the chainsaw.

A driverless car (autonomous vehicle, or AV) is programmed to kill you under certain conditions. It's clear, up front, that those conditions will be beyond your control; your self-confidence is irrelevant. Will you pay for this machine? Is the manufacturer insane to believe you would buy your own possible execution?

The AV scenario refers to the Trolley Problem—the (usually) hypothetical dilemma in which you must choose to sacrifice your own life in a traffic situation, rather than kill one or more pedestrians. Recent articles have put forward the view

© Springer Nature Switzerland AG 2019
F. Phillips, *What About the Future?*, Science, Technology and Innovation Studies,
https://doi.org/10.1007/978-3-030-26165-8_12

that the Trolley Problem inevitably will be faced by AVs, and must be considered in their programming.

I believe the chainsaw and AV scenarios describe two different psychological situations and lead to two different customer decisions. The question seems urgent; all signs point to AVs appearing in the marketplace soon and inevitably.

Scenario 1, the chainsaw, represents a common situation. We often buy products that endanger us at moments when our skill and alertness—or the skill and alertness of those around us—are not near optimum. Ordinary automobiles are a perfect example. We pay our money, we do our best, and we take our chances. This is how humans have acted for thousands of years.

Cigarettes present a situation more similar, but not identical, to AVs. Smoking-related cancer is a crapshoot; it may happen sooner, later, or not at all. The difference is that only tobacco addicts take the gamble. AVs will protect or endanger everyone who drives, rides, or walks on or near a roadway.

I am usually a technological optimist. However, I believe AVs in their currently envisioned form (i.e., on streets rather than on tracks) are unsuited for the market. Readers will also notice that much of this book deals with uncertainty reduction, a goal beloved by managers. I'm now highlighting a case where we embrace uncertainty. Why would humans embrace uncertainty? The reason that's pertinent here is, "It's the only way we can indulge our optimism bias."

A Public Broadcasting System website lets you self-test on the Trolley Problem and other moral choices. When someone—even PBS!—asks you the Trolley question, they commit a fallacy by assuming only two possible courses of action. If you, as subject, go along with the assumption, all you're doing is acting as the questioner's enabler. (Who's testing whom, really?) The best possible reply to the trolley problem is, "I don't do hypotheticals." There are always unique and extenuating conditions in every real instance of risk, including traffic risk. There is always a third choice. This makes one ask whether a machine can react appropriately to the unique conditions, to perceive and take that third or fourth option. I believe they cannot. First, static programming can prepare the machine to deal only with foreseen circumstances.

The previous sentence was just a scene-setting straw man; there will be no static programming. The central program governing the whole population of deployed AVs will "learn" and modify itself as each vehicle encounters new conditions. One problem though: the program's creators won't know what has been learned, except by watching the cars' movements. The cars' adaptive behavior will become unpredictable, and perhaps detrimental to human safety.

Okay, human drivers can be unpredictable and dangerous too. But there are additional problems with the AVs.

They are, for most of their operating time, autonomous from human control. However, they are not autonomous from each other. AVs communicate with each other and with a central computer program. This means the vendor's challenge is not just to control an individual vehicle, but to control the entire network of vehicles.

The testing of small AV networks has gone fairly well, with only one collision I recall hearing about.[1] The tests won't scale to the mass market, however, because the complexity of the system, and thus the chance of system breakdown, increases super-linearly with the number of networked vehicles.

The AV programmers may be trying to emulate flocking behavior. Each bird in a flock, following a few simple rules (e.g., "Remain about two wing-lengths away from the bird on your left"), creates highly coherent mass behavior. The problems with translating this to AV networks are (1) the rules for avoiding other AVs, pedestrians, buildings, etc., have to be more complex than for schools of fish or flocks of birds; (2) AVs will sometimes revert to the human driver's control; and (3) tests have shown the most accident-prone moments are the instants during that changeover from machine to human control. Some manufacturers are talking about AVs that will not allow a human driver to take control. I don't know whether that is more scary or less, but I don't think it would be practical for our transportation needs.

Consumers can't be sure the computer program's prime directive is to protect human life. Programmers may have given the AV network a directive to smooth traffic flow, or to maximize rider convenience. (Look at the pharmaceutical industry and the current opioid addiction crisis, for examples of how things can go wrong in terms of prime directive.) Indeed, it's my impression that articles on AVs tout life-saving as a by-product of autonomous vehicle deployment, not as the primary purpose.

Bonnefon et al. (2015) surveyed public attitudes about AVs programmed to solve the Trolley Problem using the "Utilitarian" principle:

> Although [respondents] were generally unwilling to see self-sacrifices enforced by law, they were more prepared for such legal enforcement if it applied to AVs, than if it applied to humans. Several reasons may underlie this effect: unlike humans, computers can be expected to dispassionately make utilitarian calculations in an instant; computers, unlike humans, can be expected to unerringly comply with the law, rendering moot the thorny issue of punishing non-compliers; and finally, a law requiring people to kill themselves would raise considerable ethical challenges.
>
> Even in the absence of legal enforcement, most respondents agreed that AVs should be programmed for utilitarian self-sacrifice, and to pursue the greater good rather than protect their own passenger. However, they were not as confident that AVs would be programmed that way in reality—and for a good reason: they actually wished others to cruise in utilitarian AVs, more than they wanted to buy utilitarian AVs themselves.

News items published after I began drafting this chapter suggest AVs will be rented by end users, not purchased. That is, the manufacturers are targeting the taxi and Über markets. OK, but what's really a horse of a different color is that they are pushing for urban districts in which human-driven vehicles will be prohibited. Seen

---

[1] By some accounts, that collision involved a human driver crashing into an AV. The manufacturer claimed this incident revealed nothing wrong with the AV system. I heard this kind of upside-down logic once before, when karate master Mas Oyama knocked out a bull by punching it in the head. Oyama's detractors said, "Nah, the bull was drugged." As if just anybody could punch out even a drugged bull.

from one angle, the logic is clear: in a collision between an AV and an ordinary vehicle, it's easier to blame a machine than a human driver, so the liability issues get big. Seen from another angle, it raises a fundamental question about our future: are our cities for machines, or are they for people?

In certain (admittedly, probably rare) AV crisis situations, you're dead, and you know this in advance. When you are driving, there is always hope. Where there's life, there's hope. Where there's machines, who knows? Manufacturers are rushing AVs to market without due regard for the human psychology that makes AVs unworkable.[2]

An early reader of this chapter[3] added an important dimension to the argument. He made the dark implication that the easy market for AVs will be people who—probably because they are accustomed to it—don't mind feeling helpless.

12.2   Free Markets Best?

Whenever there's a task to be done or governance to be exercised, we tend to organize for it in threes. A single power center is unworkable, as it can easily lead to dictatorship. Two is not so good either, as a disagreement can lead to indefinite and un-refereed deadlock. Three gives us "checks and balances," as all of us were taught in school.

It's not just the legislative, executive, and judiciary branches of the US government. Companies developing new products seek balance among the engineering, manufacturing, and marketing departments. Economic development rests on the "triple helix" interaction of the government, academic, and industrial sectors. Corporate governance depends on the triangle of shareholders, boards of directors, and managers.

Three seems to be a magic number!

Yet, things go wrong. Speed-to-market, essential for success in tech industries, gets torpedoed by the number 3. Marketing guys think engineers only want to make cool products that other engineers will appreciate but that no one will buy. Engineers are sure the marketers cannot possibly understand the product. Both doubt that manufacturing can actually make the thing. Except when the CEO is a brilliant, obnoxious micromanager like Steve Jobs, the conflict is unlikely to find a capable arbitrator.

To react effectively to each other's initiatives, and to be effective partners in economic development, the triple helix elements—government, academe, and industry—need to be at least semi-autonomous. Yet in America we see universities falling under the domination of business interests, even as governments cut the academic

---

[2]My lecture slides on other aspects of AV technology assessment are downloadable from Slideshare,   http://www.slideshare.net/fredphillips/the-selfdriving-car?qid=0c2ac803-13a0-4ae1-9dcb-5cb56e6d9a0b&v=&b=&from_search=1

[3]Joe Rabinovitsj.

budgets. Business lobbyists buy and sell congressmen. In other countries, the government's education ministry dominates the universities, and the great majority of companies may be SOEs, state-owned enterprises. All power imbalances reflecting the breakdown of the number 3.

Corporate scandals like Enron's happen when the CEO appoints zombie yes-men to the boards, states compete to issue the most management-friendly corporate charters, shareholder capital takes flight at the merest hiccup in quarterly returns, and boards (which are supposed to represent the shareholders) structure CEO incentives in the most perverse and stupid ways.

We organize for project management and for enterprise (and national) governance based on the theory that tripartite structures are optimal. Yet as the previous paragraphs show, three is not perfect. There are two ways checks and balances may fail. One of the three entities can dominate the other two (perhaps because one of the three cravenly failed to exercise their constitutional power, thus handing excess authority to one of the others), or two can gang up against the third. Supreme Court nominations during the Obama and Trump administrations suggest a third mode of breakdown, namely, that one or both of the executive and legislative branches so abuse their checks-and-balances power that the third entity (the Court) loses credibility.

In my youth, which was a long time ago, most economists taught that a "mixed economy" was the norm, with free markets tempered by government oversight. Since 1980 or so, a different philosophy came to dominate: that is, that government oversight (and the taxation that supports it) should be somewhere between small and none, and that a corporation's only duty is to its shareholders. Famed General Electric CEO Jack Welch initially held the latter view; later in his career he called it the "dumbest idea I ever heard."

Free markets have done a lot for our standard of living—or rather, the mixed economy did. What has happened in the recent era of extreme free markets and extreme small government?

First, we in America watched the startling rise of the Asian economies—both the "tigers" (South Korea, Japan, Singapore, and Taiwan) and mainland China. All of them became rich in this period under heavy-handed government planning and subsidies. During this time, the middle class in America has "hollowed out," with a loss of industry jobs and an increase in poverty. Other symptoms in America include poorer health and education and extreme income inequality.

Second, it has become clear that extreme free markets lead to corporate short-termism. A CEO focused solely on quarterly returns has little use for forecasting and future orientation. More importantly, short-termism can never lead to sustainability, because sustainability is a medium-term problem. (Ecological sustainability used to be a long-term problem. 2018 UN reports show it is now medium term, i.e., much more urgent than before.)

12.3 Sustainability[4]

This section makes a three-point thesis. First, that the innovation needed to achieve sustainability entails risk; risk is inseparable from sustainability. Second, that the strong link between innovation and entrepreneurship implies entrepreneurship and entrepreneurship education are keys to achieving a sustainable world. And third, that these considerations imply a change in how the UN Sustainable Development Goals, the SDGs, are to be implemented.

Stephen Hawking declared that the human race faces an existential environmental crisis within 200 years. Through pollution, overfishing, deforestation, ocean acidification, ozone depletion, etc., we are killing ourselves by killing our support environment.

We will not wait 200 years, however, to feel it; we are feeling it now. In just 2017 and 2018, hurricanes devastated Texas and Florida, a terrible tsunami hit Indonesia, and floods in East Africa and South Asia killed thousands in Sierra Leone, India, Nepal, and Bangladesh. The largest wildfires in their history endangered lives in California, Oregon, Washington, and Montana. All these disasters are partially attributed to record high temperatures due to anthropogenic climate change.

We do not now have the means to reverse these outrages. We need innovations in clean tech, in remediation technology, in energy efficiency, and so much more.

In 1987, the Brundtland Commission defined sustainability as meeting the needs of the present without compromising the ability of future generations to meet their own needs. This definition is incorrect.

Innovative technologies and products are now often provided by new ventures. Many start-ups, including those attempting environmental innovation, are risky from a market perspective—and also from technological, social, safety, and development perspectives.

How does this fact affect sustainability? Suppose I borrow money to build a green business. Green firms face the same business risks as other companies; my children might inherit a profitable business, or they might inherit nothing, my assets wiped out by the bankruptcy of my enterprise. In the latter case, I have compromised my descendants' ability to meet their own needs.

Can this be an excuse for doing nothing—for taking no risk—now? Obviously not. A green future depends on innovation. Innovation needs to be financed by debt or equity. And that implies risk. In fact, it is abjuring risk—rejecting risk—that compromises future generations.

We cannot achieve sustainability by avoiding risk or intergenerational transfers of risk. Risk is part and parcel of sustainability. Risk must be prudently embraced.

Future innovations in waste reduction, emission scrubbing, recycling, geo-engineering, ecosystem restoration, cleanup technologies, and energy efficiency will make the difference to our future.

---

[4]This section is extracted from my Kondratieff Laureate Lecture, Moscow, 2017.

The Millennium Development Goals, devised via authoritarian top-down planning, have been replaced by the United Nations' SDGs, the Sustainable Development Goals. The SDGs emerged after lengthy dialog with multiple constituencies, and consist of 17 goals and 169 targets.

How to move forward to achieve them?

An early idea was to prioritize the SDGs. This proved impractical. All the goals are important, at least to some constituencies. Prioritizing would defeat the reason for multi-constituent dialog. Constituents would demand, "Then why did you ask me in the first place?"

A later idea was to make a project plan to map which goals to work on first, second, etc. The problem with this idea lies in the uncertainty attached to innovation. Progress toward the SDGs depends on technologies that do not yet exist.

Discoveries and inventions occur at largely unpredictable intervals. The elapsed time from lab to market follows an unknown probability distribution. Government R&D funding, market acceptance of a new technology, success or failure of techno-entrepreneurs, and the emergence of dominant designs are also uncertain and risky.

No standard project management tool will suffice. We cannot temporally order the SDGs because we cannot predict the innovations, products, services, and companies that will make them possible. Then too, progress toward one goal may slow or reverse progress on one or more other goals.

We need to turn to—and devise new—planning mechanisms that embrace uncertainty, cultural-geographic diversity, and systemic connections. With that thought, I end this section abruptly, because I have only the fuzziest notion of what those mechanisms might be.

12.4   War and Other Disruptors

No matter how incisive and well-reasoned your forecast is, an event arriving "out of the blue" can disrupt it. You didn't consider the possibility of this event, because of its extremely (and I mean *extremely*) low probability, or because of its total inconceivability. Forecasters call these disruptive events "wild cards," or sometimes "black swans."

Some inconceivable events would be conceivable if we did not indulge in psychological denial. These would include asset bubbles—and their bursting—of which the 2009 mortgage crisis and the 2000 dotcom crash were great examples.

We may realistically see the coming of a war as inevitable. That's good, solid forecasting. The war itself, though, brings a host of wild cards. "Who will win" is the least of them. Business enterprises no one's ever heard of win huge defense contracts. The crisis situation opens the government budget tap; new weaponry is invented, and new countermeasures to the weapons. New methods of intelligence gathering, and new means of countersurveillance. National boundaries are redrawn. Loyal and productive citizens, suspected of sympathy with their countries of origin, are interned, their talents of no use to the war effort. Looters, at home and abroad, redistribute wealth. The winning side may act with magnanimity or with malice against the losers. Returning veterans' post-traumatic stress may make them unemployable. Soldiers form lifetime friendships and partnerships with people they would never have met were it not for the war. Many newly invented war technologies, de-weaponized, give rise to new postwar civilian industries.

Just try to foresee the (specific) results of all that! The only really predictable profits go to the arms manufacturers.

Other kinds of wild cards pop up because we don't know what we don't know, or because we do know of the possibility of an event but don't know (or underestimate) its probability.

A totally unexpected scientific breakthrough. A solar flare that disrupts radio communication on Earth. An asteroid crashing into the Earth and destroying a city. The rise of a new religion. Discovery of huge oil reserves in a third-world country. A misbehaving artificial intelligence. Wild cards and black swans!

Key Takeaways

- The prospect that autonomous vehicles (AVs) run on streets seems impractical or at least premature. Active and informed technology assessment programs should evaluate such possibilities.
- An extreme free-market economy is not prepared to deal with the sustainability problems that could spell the end of mankind. The definition of "sustainable action" needs to be changed to encompass prudent embrace of risk.
- There will always be wild cards, completely unanticipated events, which confound our predictions.

Reference

Bonnefon J-F, Shariff A, Rahwan I (2015) Autonomous vehicles need experimental ethics: are we ready for utilitarian cars? http://arxiv.org/pdf/1510.03346v1.pdf

Attitudes, Expectations, and the Future

13

More than any other time in history, mankind faces a crossroads.
One path leads to despair and utter hopelessness. The other, to total extinction.
Let us pray we have the wisdom to choose correctly.

—Woody Allen

Our preconceived views, expectations, and hopes go a long way toward creating the future, but cause us tension when the realized future fails to match our expectations. This chapter explores three examples. A large number of people seem to expect a "singularity"—a soon-to-come point in time when artificial intelligence exceeds our human intelligence—and also expect an end to Moore's Law. When masses of people change their attitudes (in this chapter's example, concerning sustainability and environment), the prospective future suddenly shifts in its nature. In marketing, branding is all about buyers' expectations. AI will bring changes in buyers' expectations. In fact, "brand" might be a thing of the past.

13.1   The Singularity

What might change if machines become smarter than us? Will we trust them to make moral, as well as rational judgments? Will we always do what they tell us to do? Will we develop an unhealthy, and maybe (as a species) maladaptive habit of depending totally on machines?

Hal Linstone and I shared a skepticism that a particular year would come when machines exceed human intelligence—or that it would matter, if it did.

We already depend on machines (though not necessarily on smart machines) to an extent that civilization would collapse if the machines disappeared. Military minds, in fact, seek countermeasures against electromagnetic pulse weapons that could knock out all electrical and electronic devices. They worry about hackers infiltrating

© Springer Nature Switzerland AG 2019
F. Phillips, *What About the Future?*, Science, Technology and Innovation Studies,
https://doi.org/10.1007/978-3-030-26165-8_13

our electrical grid. This would not change if machines were even smarter—although an escalation of smarter incursions and smarter defenses would ensue.

And as Chap. 11 implies, in our age of simultaneous bigger and smaller, complete and sudden crossovers will be no more. A crossover of machine vs. human intelligence is not a simple matter. If it happens, it will not happen all at once. A machine beat a human grandmaster at chess. It was some years later before a machine could win at Jeopardy and Go. As time goes by, machines will beat humans at additional specific tasks, but never all tasks at once.

A final argument against an impending singularity is that AI coders are, well, programmers, who are comfortable with denotative language. In general, they're not good at connotation, nuance, and context. Ah, you say, there are now robots that can simulate emotion. This is not nuance! Machines are a very long way from understanding the slight downturn of your mouth that indicates disapproval, the changed tone of your voice that signals irony, the shoulder shrug that might mean you don't know or you don't care, and the emphasis on a different word in your sentence which reverses its entire meaning.

If indeed there are multiple kinds of intelligence, machines are now mastering only one of them. Doubt the singularity.

13.2   End of Moore's Law

Don't doubt this one. No exponential trend continues forever. The size of electronic circuits will stop shrinking. Semiconductor "fabs" that cost a billion dollars will become scrap; semiconductor companies' stock prices will suffer as they write off buildings and equipment and invest in alternative technologies. What are these alternative technologies? 3-D electronic circuits, photonic circuits, and quantum computers. With luck, at least one of these will be "ready for prime time" when Moore's Law finally gives up the ghost—which semiconductor executives think will be around 2025.

13.3   Transitions

A "technology transition" occurs when an industry or a society shifts from using mostly one dominant technology to using mostly a different, presumably newer, technology. An iconic transition would be from fossil fuels to renewable energy sources.

However, we can see transitions in dominant attitudes as well as in dominant technologies.

Does it cost money to be green, or can a company make greater profits by being green? Even a decade ago, most executives thought environmental sensitivity was too expensive an option, that it would render them uncompetitive in a harsh marketplace. Today, most thoughtful CEOs understand that environmentally sustainable products and practices are keys to greater profits. Bookshop browsers now see titles like *Billion Dollar Green: Profit from the Eco Revolution* and *Green Profit on Retailing.*

What happened? How did 10 short years change attitudes and practices?

- New technologies made it easier to sort waste for resale, to burn fuel more cleanly, to use less energy, substitute safer coolants for chlorofluorocarbons, and so forth.
- More customers decided they want green products and in some cases are willing to pay more for them.
- New legislation tightened environmental regulations, making it harder to undercut competitors on price by polluting the environment.

Thirty years ago, US manufacturers thought attention to product quality was a needless cost. Inspections, design for reliability, and other quality measures stood in the way of production rates, getting enough product out the door to satisfy voracious consumer demand. Then Phil Crosby (1980) introduced the notion that "quality is free"—that companies save enough money through reduced product returns, reduced rework, and better customer loyalty to offset the cost of quality. Finally, the triumph of Japanese cars in the American market showed that customers demanded quality even if it meant backing off from our deeply held "buy American" sentiment. What changed? New technologies helped manufacturers build quality into the design and manufacturing processes, rather than simply inspect for quality at the end of an assembly line. The culture of conspicuous consumption and planned obsolescence passed, as consumers became more environmentally aware and demanding of safe, durable products. Baby boomers began to have their own children and developed a passion for serviceable, safe cribs and strollers. Ralph Nader blew the whistle on unsafe Corvairs; the Consumer Product Safety Commission banned flammable children's clothing and lead in household paints and gasoline. We became busier people, less inclined to spend our time at repair shops dealing with product defects.

Similarly, in more stable economic times when technology was changing more slowly—and these times are only 30 years behind us—efficiency was held up as the highest business virtue. But what about flexibility? Oh, flexibility is nice, the thinking went, but we don't need much of it because the business environment is not throwing us any surprises, and besides, flexibility costs money. Flexibility means finding ways to reduce machine setup time, cross-training personnel, diversifying the product line. Those costs would reduce efficiency.

Then came "mass customization," the Internet, biotech, etc. The new corporate catchphrase, and the title of the business book of the week,[1] was "adapt or die." Robots now streamline materials handling and machine setup. "Human Resource" departments, formerly dissed within firms, become the most important corporate function in Silicon Valley enterprises; every firm now wants the most resilient,

---

[1]Betts and Heinrich (2003).

creative, diverse, and versatile workforce. Efficiency is still important, but flexibility is given coequal status.

Profit and sustainability. Cost and quality. Efficiency and flexibility. In each instance, two variables were seen as trade-offs, and subsequent developments in technology, in social attitude and consumer taste, and in regulation turned the trade-off into a mutually reinforcing relationship. Environmentalism, for example, formerly an irrecoverable cost, became profitable. We call this the transition from trade-off to mutuality.

To make a simple model of the profit-environmentalism attitudinal transition, suppose that y represents profit, x represents a sustainability score (the inverse of environmental impact), and $y = x^\lambda$. As everyone recalls from high school, when $\lambda = -1$, the graph of the x, y relationship looks like the pretty curve at the bottom of Fig. 13.1; the more x, the less y, and the more y, the less x.

When $\lambda = 1$, however, the x, y graph is a straight line, like those at the top of these pictures. The more x, the more y, and the more y, the more x.

What happens when λ falls somewhere between -1 and 1? The pictures, in two rotated views, show the answer. As λ moves from -1 to 1, the x, y relationship transitions from a trade-off regime (the more x, the less y) to a mutual reinforcement regime (the more x, the more y). These intermediate points represent the gradual change of attitude.

Much would have to be added to make this a useful and measurable model of transitions. This section, with its simplified model, shows that attitudes shift, but don't shift all at once. Attitude transitions are spurred by new technologies and by seeing the views and actions of peers. Attitude transitions shape the future—not just the future of business book titles, but the future of business itself, and of the wider society.

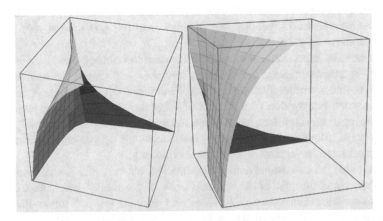

Fig. 13.1 3-D surface views of the transition. The y and x axes are on the "floor" of the "room," and the λ axis is the vertical [The graphs were made using Surface Grapher V2.0, an html utility. Model considerations owe much to discussions with Prof. Dr. Eberhard Becker of Dortmund University of Technology. We suspect transitions viewed in this way may qualify as "catastrophes" in the sense of Thom (1994)]. Source: Author

13.4   Expectations: What Do You Expect from a Consumer Product?

A research firm has just bestowed the title "world's most valuable insurance brand" on a mainland Chinese company. Other outfits issue similar announcements in diverse industries, despite that in 2014 *The Economist* made this remark about brands: "Their importance may be fading... no one agrees on how much they are worth or why."

The decline of brands: We should have seen it coming, when mass customization first began to overshadow mass production. Scholars point to info tech to explain the growing irrelevance of brands; online customer reviews and social media now substitute for the "shorthand" information packages that brands once provided.

IT is not the only force that undermines branding; the other force is confused management. But, like the movie villain who has a last-minute change of heart, IT will be the one that ultimately saves the brand.

Scholars in the 1980s declared that branding is the most powerful idea in marketing. Companies devoted the 1990s and 2000s to measuring their "brand equity."

In Germanic languages, the word brand archaically meant "burning." The brand or symbol burnt into a calf's haunch indicated ownership, identifying the calf as belonging to a ranch that was known by the same symbol. As the word began to be used in the consumer product context, it lost the "burning" connotation, at first meaning only the manufacturer's symbol or logo.

From there, the meaning expanded. Brand became a sign that customers could take confidence in the uniformity and consistent quality of the manufacturer's offerings. Later, it grew to mean an identity or image seen as a corporate asset; the value of the asset was called brand equity. Today's managers use brand to mean a unified company-wide view of quality and customer service that defines the company culture and differentiates it from others.

Unfortunately, "brand" has ceased to carry these positive connotations. Uniformity, differentiation, and ownership of offerings are things of the past.

Name-brand stores issue discount coupons. The coupons say, in small print, "Good at participating locations only." Carry the coupon to a store location, and you are told, "Sorry, we are not a participating location."

Needing help with a bank account in foreign country X, you are relieved to find the name-brand bank has a branch in your home country. You phone, and are told, "We share the brand name, but we are a different corporate entity. You must call country X."

Code-sharing means airlines are no longer airlines; they are simply travel agencies. You book a flight on airline A, only to find that the flight is "serviced by" airline B.

Department stores used to be powerfully branded. Now they are just mini-malls, leasing space to other entities. There is no Macy's shirt department. To choose a shirt you must travel from the Levi's sublease to the Tommy Bahama sublease, to... And thus no difference between Macy and Mitsukoshi.

Franchising and its growth further erode branding. Franchisors enforce less than total uniformity among franchisees who sport the same logo. Ask a question, and a franchisee may reply, "Take that up with corporate, we are just a franchise."

Outsourcing of manufacturing means a garment bearing the brand of your favorite athletic shoe company was not in fact made by that company. The actual manufacturer may not uphold the stated values of the name-brand firm.

A global branded fast-food chain deletes beef products from its menus in Hindu neighborhoods. The company then backtracks on its brand definition, saying the brand means fast, inexpensive service, rather than a consistent menu.

A hotel chain has announced it will use big data to personalize your experience. Your room's configuration, instead of being comfortably predictable, will be an algorithm's (probably faulty) idea of what you want. As soon as all major hotel chains do this, you will get something like what you want, at any chain! So much for brand differentiation.

"Place branding" is a standard tool of touristic and economic development. Ironically, it defeats itself if it succeeds. New buildings invade traditional neighborhoods. Tourists find small match between their guidebooks and the reality. Hong Kong's appearance transformed totally between 1976 and 1994, the changes including disappearance of the Aberdeen boat community, the Kowloon YMCA, and other landmarks.

All these experiences erode the meaning of brand. The "burning" question is: what will replace brand as a focus for customer confidence?

In a globally hypercompetitive economy, there is constant downward pressure on the salaries of frontline personnel, and on the budgets for training them. It was IT (cell phone videos) that broadcast the recent misdeeds of the three major US airlines—including the physical dragging of a peaceful passenger from a United flight—damaging their brands nearly to the point of no repair. Yet it was poor management policies that led to the misdeeds in the first place.

Quality of personal service will continue to be a differentiator only for premium-priced luxury brands. Confirmed passengers in the first-class cabin are never bumped from a flight! For the rest of us, quality will depend on our interface with the vendor's A.I.s (artificial intelligence algorithms).

Savvy vendors will let you talk back to the A.I., to fine-tune your interaction with it, and its idea of your preferred flavors, sizes, and services. The new brand differentiators will be the "personalities" of each vendor's A.I.[2]

---

[2]Sources for "Expectations" section: Yiu (2017), Tenet Partners (2017), The Economist (2014), Singh (2017), Metric Marketing (2017).

Key Takeaways

- The expected singularity that one not-too-far-away day artificial intelligence will outstrip human intelligence is not likely to happen as machines can never exceed humans at all tasks comprehensively. Moreover, artificial intelligences can hardy interpret connotation, nuance, and context.
- Attitude transitions that are ignited by modern technologies and peer influences change the nature of the prospective future of business and the whole society. Specifically, trade-off relationships between different variables may advance toward mutually reinforcing ones, and have done so in several domains.
- In marketing, brand will no longer be an assurance of the uniform in quality and customer services, but A.I. will.

References

Betts B, Heinrich C (2003) Adapt or die: transforming your supply chain into an adaptive business network. Wiley, Hoboken

Crosby PB (1980) Quality is free. Mentor, New York

Metric Marketing (2017) The importance of branding in your marketing. http://benchmark.metricmarketing.ca/marketing-resources/why-is-branding-important-to-marketing/, accessed 5/6/17

Singh H (2017, May 2) Big data hotels make you feel more at home. South China Morning Post, p C2

Tenet Partners (2017) Top 100 most powerful brands of 2016. https://tenetpartners.com/top100/most-powerful-brands-list.html, accessed 5/6/17

The Economist (2014, August 30) What are brands for?. http://www.economist.com/news/business/21614150-brands-are-most-valuable-assets-many-companies-possess-no-one-agrees-how-much-they, accessed 5/6/17

Thom R (1994) Structural stability and morphogenesis. Westview Press, Boulder, CO

Yiu, E (2017, May 2) China's Ping An rated top insurance brand. South China Morning Post, p B1

The Future of the Future

14

> *I do not believe in a fate that falls on men however they act;*
> *but I do believe in a fate that falls on them* unless *they act.*
> —G.K. Chesterton

We finish our discussion of the future with a few diverse questions and a few tentative answers. First, we wonder whether the short-term orientation of today's corporations will rob the human race of a future. Where there's a cause, there's effect, so we next ask, is karma always a bitch? Then we ask, is there a way to be happy anyway?

14.1 Short-Termism[1]

Though we might think only global cooperation can solve global environmental problems, globalization in its current form works against sustainability. WTO-style capital liberalization causes investment to shift quickly to the site of highest returns, irrespective of national borders. To a far greater extent than in the past, fear of disinvestment causes CEOs to strive for maximum short-term profits.

It doesn't take a genius to understand that corporate short-termism is incompatible with long-term sustainability.

When all operational inefficiencies have been squeezed out of a company, all new markets penetrated, and all M&A synergies synergized (or when the CEO thinks they have been), only three avenues to greater profitability remain: screw the employees, screw the customers, or screw the environment.

---

[1]This section was previously published as Phillips (2019). Reprinted here with permission of Elsevier.

© Springer Nature Switzerland AG 2019
F. Phillips, *What About the Future?*, Science, Technology and Innovation Studies,
https://doi.org/10.1007/978-3-030-26165-8_14

Even as sustainability scholars begin to see hypercompetition-driven profit maximizing as maladaptive, psychologists have recently identified the aforementioned CEO behavior with classical psychopathy. While 1% of the general population are psychopaths, psychologist Nathan Brooks (2017)[2] has put that figure at 21% for senior corporate executives. (Another study yields a much lower but still alarming number for executives.)

Brooks and others point out that psycho CEOs' business "successes" are short-lived, lasting only until employees realize what an SOB they're working for and depart for more life-affirming jobs. This suggests a vicious cycle of negative feedback, reinforcing short-termism.

At a 2018 workshop in Japan, I wondered why we even bother to talk about sustainability under conditions like this. I was about to challenge the speaker, Professor ZhongXiang Zhang of Tianjin University, with this question, when he uttered the phrase "Post-WTO regime."

Hmm. Serious thinkers are thinking that WTO isn't forever. Non-serious thinkers are as well, as evidenced by Donald Trump threatening to withdraw from NAFTA.

So we ditch WTO. Then what? Trump doesn't know. Deep thinkers in Switzerland are advocating transitioning the United Nations from an advisory to an administrative role. That is, to a world government.

The World Economic Forum needs to understand the profound depth of the "local control" sentiment in the USA, and the American paranoia about "elites" imposing "globalist" rules. America, it now seems, signs international agreements and then repudiates them (Paris Accords, TPP, and the Iran nuclear deal). China, too, signs and then ignores or creatively interprets them. What can remedy this?

If there is to be a post-WTO, sustainability-enhancing regime, designing it will take not just a genius, but a whole scrum of geniuses. It will be globally coupled, but not rigidly immobile. Its messages, tailored to individual cultures, will be strong but not coercive. Its financing will bring back "patient capital" while discouraging "dumb money."[3] It will reward cooperation, innovation, and data sharing. It will fail to reward venality and waste.

To close with an apt bit of doggerel,

> "The upshot is, we cannot tailor
>   policy by a single scalar,
>   unless we know the priceless price
>   of Honor, Justice, Pride and Vice.
>   This means a crisis is arising
>   for simple-minded maximizing."
>   —Kenneth Boulding

---

[2]Serial CEO Margaret Heffernan makes a reasonable rebuttal of the "the executive suite disproportionately attracts psychopaths" theory, at http://www.cbsnews.com/news/are-ceos-psychopaths/

[3]"Dumb money" = desperate investments in questionable vehicles, in times of low or zero interest rates.

14.2 Creating Karma?

I saw a Facebook post by someone with a name similar to that of an old friend, whom I had not heard from (or of) for nearly 40 years. It led me to Google my friend from long ago. I was shocked to learn she had passed away some years back, after years of decline following injuries in a car crash. When we were teenagers, she wanted to be a nurse. I told her being female would not hold her back (even though this was the 1960s); someone with her smarts could easily become an MD. She did indeed become a doctor.

If she had not gone to med school, would she have been in that place at the exact time the truck plowed into her?

I am skeptical of the Buddhist monastic injunction against creating karma. If I am sitting in the zendo, and as a result I am not out there to help the proverbial little old lady across the street, a bus might flatten her. If I do save her from injury by helping her across the street, maybe she'll poison her husband. Either way, I've created karma.

My friend's obit said she was known for her compassion and as a doctor had helped a great many people. She was never on social media; she did good quietly—my teacher Koichi Tohei would have called this *intoku,* admirably doing good in secret. All I can do is be proud of my friend, and reflect that such a reputation is the best most of us can hope for in a lifetime.

Every decision will have good and bad consequences.

Buddhists are wise to emphasize the karma principle, but less wise to believe karma created in "past lives" catches up with you in this life. Serial reincarnation presupposes one's essence lives on in some other dimension while waiting to be reborn in this one. So far, so good: That's the only way it could work. But to suppose that time, in that other dimension, proceeds in the same way it does in ours is an unwarranted assumption, indeed a ridiculous one. Does "after" in that world mean the same as "after" in this world? No reason to think so.

So let's float an alternative idea: You are already reincarnated, in many places, in this world, right now. Personality is a product of genetics and environment. Carl Jung said there are only so many dramas (environments) in this world, and so many roles in them. The number of human genotypes is huge, but I would guess it is comparable to, or perhaps less than, the 7.5 billion that is Earth's human population. Thus... several people just like you are doing exactly what you're doing, in analogous dramas, elsewhere on the planet.

I'm surely jumping off a doctrinal cliff by putting forth that theory. Yet, it seems in closer accord with the idea that we are one with the universe. Anyway, it's just for fun. Zen Buddhists don't bother with the reincarnation shtick in any case; it's more a preoccupation of the Mahayana. Karma is created and reaped in this lifetime.

Buddhists are accused of bad taste and insensitivity, for applying the karma principle, for example, to the Holocaust. The six million, they say, did awful things—and probably weren't Jewish, or Gypsy, or gay—in an earlier life. To modern ears, this sounds like blaming the victims. It lets those who could have intervened off the hook. Dispensing with the notion of serial reincarnation eliminates this sticky point, but preserves the valuable principle of karma.

14.3 Happiness

It's banal to mention that technology is a two-edged sword. That it solves practical problems and creates new ones. That it makes our lives more comfortable and more complex, and stresses and at the same time sustains our social relationships. In this chapter, we'll go beyond these commonplaces to explore two lesser-known aspects of tech's dark side: inequality and unhappiness. Will the dark side prevail? Maybe, but we'll see glimmers of hope for the team of truth and goodness.

14.3.1 The Growing Gap

Technology-enabled globalization meant investment capital could freely roam the world, looking for the greatest returns. Investors who could buy information about the location of the greatest return could invest accordingly. Wealth built on itself, turning itself into extreme wealth.

Then too, more advanced business machines require more educated operators. Education is becoming more expensive, and students are taking on unsustainable debt to get it. Richer school districts have better teachers and equipment, and the gap in student test scores between rich and poor districts is growing.

Thus, technology is one cause of today's extreme wealth inequality.[4]

In earlier times, technological advance equated with new equipment for mining and manufacturing—what the accountants call capital equipment. By definition, these innovations were available only to those who had capital, lots of it. And there's no question that inventors were bankrolled by capitalists who wanted newer ways to gain leverage over the laboring masses.

Some exceptions proved this rule. The mule- or ox-drawn plow was at one time a new technology, and it enriched small farmers, at an affordable investment cost. Later, the sewing machine allowed small entrepreneurs to start tailor shops—but also enabled more ambitious textile entrepreneurs to perpetrate sweatshop manufacture.

It was the recent genius of Japanese and Silicon Valley entrepreneurs to follow the sewing machine example, making productive equipment (notably, PCs, digital

---

[4]Alan Blinder, former vice chairman of the US Federal Reserve, writing in the *Wall Street Journal,* lays the blame for income inequality on technology. Other authors claim inequality is cyclical, not structural. A United Nations University study (Singh and Dhumale 2000) concludes globalization and technology are not the primary drivers of inequality, though authors admit technology is a significant contributor. Their paper addresses income inequality, not wealth inequality.

Tellingly, "San Francisco, the heart of the tech industry, now has the fastest-growing income inequality in the country, a gap on par with Rwanda's." http://recode.net/2014/05/31/tech-titans-on-income-inequality-and-their-stingy-stingy-industry/

The poor idiot economists think inequality just means there's a bigger gap between the rich and the rest of us. Not true: Extreme inequality changes everything, including the structure of society and industries. The new structures create latent demand for still more technological innovation (see "The Circle of Innovation" back in Chap. 6), and the cycle continues.

media equipment like mixers and DVD duplicators, and smartphones) affordable for everyone. The equipment enabled hundreds of thousands of new small businesses. If not for the ability of ordinary people to save money and make money using their digital devices, today's near-intolerable income inequality in the USA would be ever so much worse.

Not enough attention is given to this countervailing trend, even though the impact of these democratizing innovations on wealth inequality (as opposed to income inequality) so far appears to be small.[5] The trend (a light saber wielded against the dark side!) is made possible by Moore's Law and related cost trajectories: transistor density on integrated circuits continues to double every 20 months or so at constant cost; the price of DNA sequencing is dropping even faster; and solar power $/kwh ditto.

Bravo, and let's see more of it!

14.3.2 Economics Meet Politics

It's worth remembering that the science fiction we read as youngsters—even the works of libertarian authors like Heinlein and Campbell—presumed some kind of mild socialism, a basic dole, in the high-tech futures they painted. (Heinlein, trained in math and engineering, was a better economist than the economists!) After all, with big economic surpluses from robotic production, why not?

Some technopreneur millionaires in our own day rail against the taxes that would support this future, preferring to give to selected charities. Others, suffering under the delusion that being good at making money automatically means they're good at giving it away, only give to charities they own or control. The situation has become so dangerous that even *The Economist* magazine, a bastion of conservative thought, is now advocating government-led redistribution.

Big corporations, of course, offshore their profits in order not to pay US taxes, and sit on piles of cash that in a sensible world could be put to productive or humanitarian use. Their CEOs gamble that violent social revolution will wait until after they're retired or dead.

14.3.3 Are We Happy Yet?

We'd be silly to expect those holding the short end of the inequality stick to be happy campers. But let's put aside inequality, just for the moment, and look at other effects of technology on happiness.

New technologies cause social change. This can mean change in the structure of society—change in its roles and institutions, which I addressed in the journal

---

[5]See for example iPhone: The Affordably Luxurious Global Accessory. http://www.thelowdownblog.com/2014_08_10_archive.html

Foresight[6]—or it can mean change in the outcomes of social organization. The most wanted outcome, naturally, is happiness. As a general, top-level syllogism, we know people dislike change. Tech advances create change. Therefore, we should expect tech advance to reduce happiness, not increase it, at least in the short run.

Can it be otherwise? There must be threshold effects. At the end of the Korean War nearly 70 years ago, South Korea was the poorest country in Asia. Now, it is the richest, or nearly so. Its increased wealth came from technology-based industries, as all Koreans are aware. Poverty, combined with postwar hardships, was not a recipe for happiness. There was an upward bump in the general happiness when people had enough to feed their children and themselves. And another when medical technology reduced infant mortality. There was still another bump, according to one of my informants, when the country reached the stage where people could feed their families by doing honest work, without stealing from each other and from tourists.

That might have been the end of the upward trend in happiness, though. Today, the pace of tech change in South Korea remains breathtaking, far beyond what's common in the USA. And the number of Koreans entering hospital with stress-related illnesses is also growing proportionally faster than in the USA. This is especially striking because admitting to mental illness has always carried more stigma in Korea than in America.[7]

The studies[8] purporting to show a positive relationship between technology and happiness seem methodologically flawed, either by not defining the two variables precisely, or by measuring them with sketchy instruments, or by not recognizing the threshold effect. One study found a positive relationship but admitted there were "decreasing returns to scale," that is, that doubling the technology did not double the happiness. We knew that already, of course, having discussed concave utility functions (Chap. 8), and given that no one has yet invented Woody Allen's orgasmatron. (In the latter regard, though, let's acknowledge the new Pew Research report that predicts sexbots will be "commonplace" by 2025.)

Then too, based on a single question about subjective feelings of well-being, "The Pennsylvania Amish [who use minimal technology], when asked how much they agree with the statement 'You are satisfied with your life (using a scale of 1–10),'

---

[6]Phillips (2014).

[7]"Since the 1950s, reports of major depression [in the USA] have increased tenfold, and while much of that increase undoubtedly represents a new willingness to diagnose mental illness, there's a general consensus among mental-health experts that it also reflects a real development." http://www.technologyreview.com/review/403558/technology-and-happiness/

You will no doubt remind me that there are (Wall) street thugs in America who still enjoy stealing from us, and you'll guess correctly that stealing is not extinct in Korea either. Moreover, new technologies make newer and more interesting opportunities for stealing, like identity theft and other Internet cons. However, perpetrators are a smaller fraction of the population than in decades past, and this is a digression, anyway.

[8]Like http://www.bbc.co.uk/news/10108551 and http://readwrite.com/2014/04/22/relationship-technology-happiness-countries

turn out to be as happy as the members of the Forbes 400," according to James Surowiecki in *MIT Technology Review*.

There is no convincing evidence that technological innovation directly moves people upward through Maslow's hierarchy of needs.[9] As additional countries emulate Bhutan's measurement of "gross national happiness," we might eventually reach an evidence-based conclusion on that matter.

Meanwhile, researchers at MIT and Hitachi are combining sensor technology with "quantified self" stuff, to directly interface with the physiological indicators of happiness:

> By monitoring and analyzing a person's sleep patterns, exercise and dietary habits, and vital statistics like body temperature, blood pressure, and heart rate, [sensors] can pinpoint trouble spots in the person's daily routine and then suggest modifications that measurably improve that individual's outlook and well-being. (Yano et al. 2012)

What's that you say? A machine cannot make me happy, only I can make myself happy? Well, tech is always a two-edged sword, generating its own abuses, and the sensor researchers' findings may or may not make us happy. At least, they'll make us think more deeply about what happiness is, and where it comes from.

Television pundit Jon Stewart has said, "How do you know what is the right path to choose to get the result that you desire? The honest answer is this: You won't. And accepting that greatly eases the anxiety of your life experience." Or as musician Bobby McFerrin put it even more simply, "Don't worry, be happy." This approach may strike some as irresponsible or anti-intellectual. For others, it might be the best path. To each his own—but before you choose, reread the Chesterton quote at the top of this chapter.

14.4 Ingredients

A member of the Danish parliament has painted a scenario for the World Economic Forum provocatively titled "Welcome to 2030: I own nothing, have no privacy, and life has never been better." Her scenario—in which she depends on the online sharing economy, and trades personal data for goods and services—has the Copenhagen city government driving this Smart City as a Service (SCaaS) model. Cities may be more attuned and responsive to local needs than national governments, but my European correspondents say most European cities do not have the authority to act in this way. We shall see.

---

[9]Certainly, gadgets can make people smile. My daughter goes so far as to say, "Maslow forgot to mention consumer electronics!" Tech entrepreneurs have found limits to selling efficiency. Rishi Mandal, cofounder of Sosh, says, "The next question is, how do you accomplish tasks while creating a smile?"

In any case, our Parliamentarian is envisioning a shift in values—regarding her personal privacy and private property—and a dependence on information technology.

In his book, *The Wizard and the Prophet: Two Remarkable Scientists and Their Dueling Visions to Shape Tomorrow's World,* Charles C. Mann profiles two men having "radically divergent views on what strategies would save humanity" and whose views "still inform environmental action today." One of the men is Norman Borlaug, who devised and drove the Green Revolution, using technology to increase crop yields, thus saving hundreds of millions of lives since the middle of the last century, and winning the Nobel Prize for this work.

On the other hand, William Vogt, the second man to be profiled by Mann, urged frugal and green lifestyles. He "believed that self-control, not science, would save us."

Now that you have read *What About the Future*, and understood its emphasis on the interplay between social change and technological change, you know that *both* Borlaug and Vogt are correct—though either view might be more efficacious in a particular place at a particular time.

Technological innovation *and* personal and social convictions (yes, and political and religious ones, too) are the ingredients of the future.

We have come to the end of the book. You now have another set of ingredients also, namely, those that will help you form a philosophy of the future.

My students (and other people!) say weirdly redundant things like "plan ahead," "forecast the future," "advance warning," and "future prospects." Clearly, they need a sharper comprehension of the future; this book provides it.

I check in at the airport, and the clerk asks, "What is your final destination?" I think, "Buddy, I know my final destination, I just hope it's not today!" Clearly, the clerk and I tend are using different timescales to think about the future. This book shows that reconciling team members' varied concepts of the future is an essential first step in project planning.

From here, you will graduate to other books that set forth the detailed techniques of planning, forecasting, foresight, and decision. Here's to your constructive and happy future.

Key Takeaways

- The race for short-term profits of corporations caused by globalization is damaging long-term sustainability. Prescribing a remedy to this problem will require plenty of geniuses gathered together to come up with a regime that values Honor, Justice, Pride, and Vice.
- We create karma with every action we take, and we will garner karma in this lifetime.
- Technology is a two-edged sword that leads to intense wealth inequality but also can prevent the inequality from worsening. At the same time, technological advances make people happier, but also cause more stress-related diseases.

Technological innovation together with social changes are the ingredients of the future. How we use those ingredients to become happier is our decision.

References

Brooks N (2017) Understanding the manifestation of psychopathic personality characteristics across populations. PhD thesis, Bond University, Faculty of Society and Design

Phillips F (2014) Meta-measures for technology and environment. Foresight 16(5):410–431

Phillips F (2019) The globalization paradox. Technol Forecast Soc Chang 143:319–320

Singh A, Dhumale R (2000) Globalization, technology, and income inequality: a critical analysis. UNU

Yano K, Lyubomirsky S, Chancellor J (2012, November 28) Can technology make you happy? IEEE Spectrum, https://spectrum.ieee.org/at-work/innovation/can-technology-make-you-happy

Printed in the United States
By Bookmasters